THE COMPLETE GUIDE TO

CONTRACTING YOUR HOME

THE COMPLETE GUIDE TO

CONTRACTING YOUR HOME

A Step-by-Step Method for Managing Home Construction

Dave McGuerty
and Kent Lester

Betterway Publications, Inc.
White Hall, Virginia

Published by Betterway Publications, Inc.
White Hall, VA 22987

Every precaution has been taken in preparing *The Complete Guide to Contracting Your Home* to make your home construction project as safe and successful as possible. However, neither Betterway Publications nor the authors assume any responsibility for any damages incurred in conjunction with the use of this manual.

Library of Congress Cataloging-in-Publication Data

McGuerty, Dave,

    The complete guide to contracting your home.

    Includes index.
    1. House construction--Amateurs' manuals.
I. Lester, Kent,   . II. Title.
TH4815.M27 1986   690'.83   85-31789
ISBN 0-932620-57-4 (pbk.)

Printed in the United States of America
20 19 18 17 16

Dedicated to you:

THE HOMEBUILDER

That rare breed of individual with determination,
perseverance, demand for quality,
and the desire to save lots of money.

ACKNOWLEDGMENTS

To Thomas Harrison
and
The Georgia Institute of Homebuilding
for their contribution and support.

Many illustrations in this edition appear courtesy of the
United States Department of Housing and Urban Development
and the United States Department of Agriculture.
We wish to thank many of its members for being so helpful.

Contents

Introduction

Welcome to the exciting world of homebuilding. Whether you are a homebuilder or an individual planning to build your own home, this manual will be a great step in helping you to achieve your goal.

INTRODUCTION TO HOMEBUILDING

Homes have existed as long as man himself. Over centuries, homes have evolved from simple dwellings to complex structures combining state of the art skills and materials from many industries. Such rapid advancement has made the task of building a home increasingly difficult and complex. What the residential builder needs is a simple and complete method covering all of the activities involved in completing a construction project from beginning to end. That's what this book is all about.

WHO CAN BE A HOMEBUILDER?

A homebuilder is anyone who wants to be one. Many states have no examinations or licensing requirements for builders. Check with local building authorities in your area to determine if you must meet any special requirements. The Home builder, or contractor, is the person in charge of overseeing the entire construction project, coordinating the efforts of professional sub-contractors and labor, and managing materials. The term "Homebuilder" is somewhat of a misnomer since a Homebuilder normally does not even have to raise a hammer to build a home. His efforts are best concentrated on managing the project. A professional Homebuilder is anyone who builds for a living.

THE RESIDENTIAL CONSTRUCTION METHOD

Many would-be homebuilders are discouraged from attempting to build their own homes because of fear due to the "mystique" of modern construction practices and the profusion of available reference materials. Although many steps covered in this book provide you with information on what is happening at a specific point in time, you are usually not the person

doing it. The steps you choose to perform are up to you. This book provides concise step-by-step instructions for residential construction Building a home requires a sound knowledge of contracting. Contracting is the process of coordinating the efforts of skilled tradesmen (carpenters, masons, electricians, plumbers, etc...) and materials in an efficient manner. This book will show you how to work with subcontractors and material suppliers to become a successful Homebuilder, whether you plan to build one house or a hundred.

MANAGEMENT OBJECTIVES

- SAVE TIME
- SAVE MONEY
- REDUCE FRUSTRATION

SAVE TIME

This book provides you with a complete detailed master plan that will help keep your project organized by dividing the work into discrete, manageable tasks. Efficiency will be gained by only doing necessary tasks - and doing them only once.

SAVE MONEY

This book can easily help save you from 25 to 40 percent on the home you build over a comparable home purchased through a real estate agent. Here are the major ways:

-Real Estate Commission	up to	7%
-Builder Markup	up to	20%
-Material savings	up to	4%
-Cost saving construction	up to	2%
-Doing work yourself	up to	9%

Total	up to	42%
		======

This book will also provide you with a wealth of instantly usable knowledge equivalent to many other books on construction. This book can also save you from numerous, costly pitfalls associated with residential construction. After all, learning from bad experiences doesn't mean they have to be your own!

REDUCE FRUSTRATION
This book will:

■ EQUIP you with the knowledge to effectively manage a residential construction project.

■ HELP you avoid major problems and pitfalls.

■ PROVIDE you with the benefit of years of experience.

■ ADVISE you when to avoid tackling tasks which are best left to specialists.

■ SHOW you how to specify and inspect the work performed by others.

HOW THIS BOOK DIFFERS FROM OTHER CONSTRUCTION MANUALS

There are more than enough books on the market covering construction techniques such as how to frame a home, lay bricks, mix concrete, cut beveled miter joints, etc. However, most fail to cover the important, practical day-to-day details of managing a construction project. Remember, you are not building the home per se; you are managing the building of the home. Also, many other publications fail to provide the necessary forms and sequenced, step-by-step instructions helpful in guiding you through the day to day details of a construction project.

OVERVIEW
This *COMPLETE GUIDE TO CONTRACTING YOUR HOME* is comprised of two major sections:

■ CONCEPTS
■ HOMEBUILDING PROCESS

CONCEPTS
Certain aspects of construction do not deal with specific construction steps, but are important ingredients to understanding successful home-building. Concepts such as lot and plan selection, financial and legal issues, dealing with members of the construction industry and other topics are important in gaining a proper orientation to this environment. Much of the CONCEPTS material will be helpful throughout the PROJECT MANAGEMENT section of this book.

HOMEBUILDING PROCESS
Residential construction involves the orderly coordination of some two dozen specialty trades each with their own special instructions, specifications, inspection requirements, detailed steps and special terms. The home building section will cover each trade one at a time in the order in which the project normally proceeds.

HOW TO USE THIS BOOK
This book is designed to be easy to use and straight forward. It has been carefully sequenced to provide you with the right information at the right time.

Before beginning your project:

■ READ this book thoroughly.

■ FAMILIARIZE yourself with all material in the HOMEBUILDING PROCESS section.

During your project:

■ FOLLOW the STEPS outlined in the HOMEBUILDING PROCESS section. Certain steps may not pertain to your project. For example, if you are building on a slab foundation, excavating a basement will not be applicable. However, be very careful not to skip steps that do apply to your project. Some common sense and judgment is obviously essential in this respect. Interpretation of the execution of many steps will naturally vary from project to project. DO NOT EVEN THINK OF STARTING YOUR PROJECT UNTIL YOU ARE FAMILIAR WITH THE STEPS COVERED IN THIS BOOK.

■ REFER to the sample specifications for ideas on what to include in your specifications when dealing with tradesmen. These are primarily intended to remind you of items easily overlooked and to illustrate the level of detail you should use.

■ REFER to the INSPECTION checklists for ideas on how and what to inspect to help insure you are getting good workmanship as your project progresses. It may take ten years to learn how to lay bricks properly, but only ten minutes to check the work for quality craftsmanship.

■ USE the FORMS provided when applicable. Most of the following forms are included in the Appendices:

» House plan evaluation checklist
» Lot evaluation checklist
» Sample Building Contract
» Cost Estimate sheets
» Purchase Orders
» Subcontractor Specification Form
» Change Order

EXCUSES FOR NOT BUILDING

If you decide not to contract your own home, here are a few excuses you can use:

■ I have plenty of money. I don't need to save money by contracting myself.

■ I'm not particular about getting exactly what I want in a home, representing the single largest financial commitment of my life.

■ I don't know how.

■ It seems too complicated - I don't know where to begin.

■ I can't even hammer a nail straight.

■ It's not worth the trouble.

■ I don't have the time.

■ It's not my bag.

The last excuse is the only valid one. Building isn't for everyone. Many will use one of the above excuses or custom make their own. To use an excuse is to rationalize why one pays "retail" for a home. Sadly enough for some, the difference between owning a home and not, lies in the money saved by contracting.

The second most valid excuse is:

"I DON'T HAVE THE TIME"

Most of us who can afford to contemplate building a home have a full-time job. So how can one build a home in the evenings and on week-ends? This is resolved in the "How To Be A Builder" section.

HELPFUL PERIODICALS

The magazines listed below are some of many you may wish to use as a source of ideas:

Fine Homebuilding
Better Homes & Gardens
New Shelter
Architectural Digest
Southern Homes
Southern Living
Country Living
House & Garden
Fine Homes
Town & Country
Southern Accents
Home

ADDITIONAL HELPFUL MATERIAL

In conjunction with your manual, several other documents will provide you with a wealth of information, while minimizing your expenditures on reference material. They are:

Your local building code - These are the rules you will have to abide by while building in your area. Obtainable from your county building inspector.

THE ESTIMATOR - A paperback full of tables and graphs to help you estimate the quantity of each building material you will use. Available from THE HOME BUILDING PLAN SERVICE, 2235 N.E. Sandy Boulevard. Portland, Oregon 97232.

THE LABORATOR - A counterpart to "The Estimator". This book shows time in hours and minutes, required to perform various construction activities. Same publisher as above.

MODERN CARPENTRY - A general purpose text with over 1400 excellent illustrations covering the actual techniques of construction. By Willis H. Wagner. 1973. 480 Pages. THE GOODHEART-WILLCOX CO., INC. South Holland, Illinois. ISBN 0-87006-208-5.

MINIMUM PROPERTY STANDARDS FOR SINGLE FAMILY HOUSING - HUD guidelines that FHA and VA officials use to evaluate homes. May 1979. 325 pages. U.S. Department of Housing and Urban Development. Available from the Superintendent of Documents, U.S. Government Printing Office, Washington, D.C. 20402. Free or nominal cost.

MANUAL OF ACCEPTABLE PRACTICES - Extensive manual of specifications covering most aspects of residential construction. Document number 4390.1 . U.S. Department of Housing and Urban Development. 1973. 420 pages. Washington, D.C. Available from the Superintendent of Documents, U.S. Government Printing Office, Washington, D.C. 20402. Free or nominal cost.

SECTION I: CONCEPTS

1. How to Be a Builder

A lot of people incorrectly believe that builders must know everything about building a house and must pass several tests to become licensed. Many states do not require small homebuilders to be licensed, and in states that do, building your own home is exempt from licensing requirements.

Few working people can afford the time to build their house themselves by hand. This book is directed at the person wanting to function as the CONTRACTOR. The contractor can be thought of as manager of the project, making sure that sub contractors and materials show up at the site when needed. Leave the difficult and tedious tasks to the experts: They have spent years learning their trade.

Homebuilding may challenge your abilities in ways yet untested. It will test your ability to:

- Get people to do what you want and when you want
- Be fair, yet tough
- Save money in creative ways
- Manage an exciting, major project. You are the BOSS.

Before you embark on this major project, you must become educated. Just as important, after you have studied this manual, you must decide whether you can handle all the details, people, surprises, insanity and periodic frustration associated with homebuilding. Building is not for the squeamish. This book will not necessarily make homebuilding a bed of roses for you. Homebuilding favors hard work, perseverance, guts, the ability to "horse trade", common sense and anything else you can muster to get the job done. To get yourself on the right path to being a builder, get your mind set for doing two things right off:

- Act like a Builder

- Keep Good Records

- Exercise common sense

THINK LIKE A BUILDER

Remember: Unless special licensing is required in your area, you are just as much a builder as the next guy. But check with your county Building Inspection Department for a copy of the local building code and information on licensing.

Get comfortable using building terms and present yourself in a self confident and professional manner. Trade-specific terms are covered in the respective sections in the HOME-BUILDING PROCESS SECTION. The respect of your subs, labor and suppliers is essential to your effectiveness as a builder. The better you relate to those you work with, the easier your job will be. As good practice, get in your jeans, construction boots and your favorite plaid shirt and walk around a few construction sites. Get some mud on your shoes. This is a dress rehearsal. What you see is probably similar to what you'll have when you break ground. Get used to the smell of lumber, lots of mud and nails by the box. Talk to subs and imagine they're working for you. Figure out what makes them tick; and keep telling yourself you're a builder!

How you are perceived by others in this industry can make a difference. Much of a contractor's job is human relations, so acting like a builder will help you in getting along with others involved with your project. Order a set of business cards. It is amazing how many doors a simple business card will open. The business card will help convince suppliers that you are indeed a "professional". The only time you should wear a suit while working on this project is when you're at the bank polishing apples.

KEEP GOOD RECORDS

If this manual accomplishes anything, it should be to emphasize the need for keeping track of costs and expenses. You will be at a distinct advantage over the common builder if you keep good records. Some of the primary records are:

COST TAKE-OFF - The single most important tool you can use to save money on your project is the cost takeoff. The cost takeoff will tell you approximately what your house will cost before starting the project. It also gives you that checklist of materials necessary to make sure that all essential materials are ordered at the proper time. The takeoff has a further advantage of giving you a cost goal to strive for when purchasing materials. When bargaining with materials suppliers don't hesitate to point out that "I've got to stay within my cost take-off budget!".

PURCHASE ORDERS - Make sure to use purchase orders when ordering materials. This will eliminate possible future confusion over whether certain items were ordered and from whom. Purchase orders and invoices are ESSENTIAL when figuring your house's cost basis for the IRS.

INVOICES - Most of the checks you write will be initiated by invoices. It is your responsibility to make sure they are accurate and that you pay them on time; especially if discounts apply. Match your purchase orders to invoices to control payments.

RECEIPTS AND CANCELED CHECKS - Lots of money is going to flow out on invoices. Keep them in a book. You'll need to compare them against purchase orders to avoid paying for things you didn't order and paying for things twice. Compare what was delivered with the invoice. Keep all proofs of payments together in an organized manner. Try to get a receipt - not just a canceled check. Your canceled check may take a month to get back to you as proof of payment.

PHOTOGRAPHS - Consider taking photographs of your home at various times during construction. This allows you to visually see changes take place and record it. Use a tripod and mark three spots on the ground. This way, you will always be taking a picture from exactly the same location.

CONTRACTS - Nothing will make or break the project like being able or unable to lay your hands on a good, sound written contract when a dispute arises. A man's memory is only as good as the paper it's written on. Contracts and their related specifications can almost never be too detailed. You can specify anything you want; but if you don't specify it, don't count on it being done.

WORKMANS COMPENSATION RECORDS - You must keep track of which subs have their own workman's compensation policy. For all those who don't you should retain approxi-

mately six percent of the total charges. For subs with their own Workmans Compensation, you should make note of tneir policy number and expiration date.

A SMALL PILE OF BANK LOAN PAPERS - These papers will vary from one lending institution to another, but will include all or some of the following:

- Loan documents

- Draw Schedules

- Funds drawn

- Original takeoff

- Surveys and plats

- Title search and insurance

THE BUILDER'S ATTITUDE

Your attitude and that of others working on your project is important. You may not always be able to see it, but a poor attitude can cause:

■ Delays
■ Liens
■ Poor workmanship
■ Additional costs due to carelessness
■ Lots of frustration and headaches

Undertaking such a bold project takes a special frame of mind. And once you break ground, there is no turning back. Here are some important attitudes you'll need to build successfully:

■ PERSEVERE. Don't let anything get you down. When problems arise; tackle them immediately. Keep the project moving forward as close to schedule as possible. Time is money and you are responsible.

■ BE FIRM. When you have a dispute with a sub over sloppy work or a high bid, stick to your guns. You are the BOSS. Don't let anyone talk you up against a wall. Remember the golden rule: "He who has the gold makes the rules".

■ DON'T BE A PERFECTIONIST. You want the work done on your home to be of high quality, but don't be upset every time you see a bent nail; (provided you don't see too many of them). Chances are, the flaws you see in your home will not be noticeable by others when completed.

■ BE FRUGAL. Always look for new and creative ways to save money on anything pertaining to the project. Small savings can really add up in a project of this size.

■ BE THOROUGH. Think your ideas through and be meticulous. Pay attention to detail. Have the desire to keep detailed records organized and up to date.

■ DON'T GET MAD. Avoid chiding your subs on the job as this will reduce their desire to cooperate with you.

■ DON'T WORRY TOO MUCH. Instead of worrying, do something about the problem.

THE BUILDER'S RESPONSIBILITIES

The laborers and sub-contractors on your project are responsible for performing quality work and providing materials as specified and when specified by written contracts. That is their single most important responsibility. As the contractor, your responsibilities include:

■ PROVIDE clear, detailed specifications for those who work for you.

■ PAY laborers and sub-contractors promptly. Paying by check always gives you a receipt.

■ SCHEDULE material to arrive when or just before needed.

■ SCHEDULE labor and subs to work when needed.

■ MINIMIZE costs while maximizing quality.

■ PROVIDE a safe site at which to work.

■ MAINTAIN control of project and subs.

■ INSPECT work as it proceeds and is completed.

■ MAINTAIN accurate, up-to-date records on all project matters such as:

- Insurance
- Drawings and plans
- Inspections
- Bank papers
- Purchase orders and invoice
- Legal documents such as variances, permits and licenses
- Payments, receipts, bank documents and other financial matters.

HOW TO BE A PART-TIME BUILDER

Most people have a full-time job and cannot take six months off to construct a house. How can you have your cake and eat it too? There are at least two good options available.

OPTION I - HIRE A SUPERVISOR.
Check the want ads for retired carpenters or builders who would like some light work in the construction industry. They have the time and experience to do things right, and chances are, they've been in the building industry most of their lives. Even if they are too old to do physical work, they can spot good work - and bad. This is also a great source for subs and the occasional use of a pickup truck.

Arrangements can probably be made on an hourly, daily or even monthly basis. They can be around at three o'clock in the afternoon when the concrete is about to be poured or a load of bricks needs to be delivered when you are at work or out of town. These individuals can play an important part in your project. Without them, you may not be able to build at all. If this route is chosen, involve the individual early in the process, when designs and project are being planned. This will allow you time to determine the individual's credibility, skill and honesty. If you would like more control over your project, consider getting a pager, so that you can reach the supervisor at all times.

Most subs can do work unsupervised when not being paid by the hour. Your "supervisor" only needs to confirm that they are doing the proper work, answer questions and assist in inspecting the work. He's there to keep the job rolling along and answer questions that would otherwise bring work to a halt.

OPTION II - PAY A BUILDER FOR HIS ASSISTANCE.

Another option is to pay another builder to help run your project. This allows you:

- Use of his subs. This saves you the time and hassle of finding reputable subs. In return, let him put up his building sign at your site.

- Assistance during your work hours. If you can be at the site before and after work, this can be a helpful bridge during those unattended hours.

- Assistance in securing a construction loan.

In return there may be a number of favors you can do for the builder when you don't have anything to do at your site. Also, your site may attract another prospective buyer for your builder. Let him keep a copy of your plans if he likes them.

THE HOMEBUILDING PROCESS

Regardless of what size home you are building or where you build, there are certain inevitable characteristics your project will have.

YOUR PROJECT WILL PROCEED WITH DECREASING DEGREES OF RISK.

For example, the first steps such as digging the foundation and pouring footings will have more impact if done improperly then if a door is hung to swing in the wrong direction. Hence, your utmost care should be taken at the beginning of the project when every step is critical - particularly through the framing stage.

YOUR PROJECT WILL PROCEED ERRATICALLY.

As the result of unforeseen circumstances, many weeks may pass with no progress on the project. Material shortages, poor weather, or subcontractor scheduling can all contribute to delays. Then for no apparent reason, three different subs and all your materials will show up at the site at one time.

YOU WILL MAKE CHANGES AND UNPLANNED, IMPROMPTU DECISIONS.

You will probably change the position of a door, window, wall, or major appliance. You may put in a dormer, a skylight or a screen porch. A prefab fireplace may be substituted for a masonry one at the last minute. There is no limit to the things that can change once a project is started. Don't let this frighten you. Details not necessarily shown on your plans will unfold before your very eyes. These trivial issues are normal and always occur when an idea on paper is converted to reality.

YOU WILL SPEND LOTS OF MONEY AND USE IT AS A TOOL.

Until you have paid a sub you have his undivided attention. If he doesn't fix those squeaky stairs, that leaky pipe, or that crooked window, you simply won't pay him. Once you have paid too much, too early, you have killed their incentive to return and you have also failed to use a builder's most important tool.

YOU WILL MAKE THE DIFFERENCE.

Small jobs will appear with no one assigned to do them. Replacing a bad piece of plywood or sturdying up the floor with a few nails - the list is endless. You make the difference. You may sacrifice a lazy morning in bed to pick up some brick, test out paint samples, or get a free load of slate somewhere. If you are determined to get the job done, you will get up at five AM if that's what it takes. You will crawl in bed some nights aching - but knowing you did what had to be done to make a lasting contribution to a structure that may well last a hundred

COMMON PROBLEMS FOR BUILDERS

Being a builder means handling problems. Although every project is different and has its peculiar quirks, there are several problems that occur to some degree on just about all construction projects. Although you may not be able to avoid these problems altogether, it may be helpful for you to know what some of them are now so that you take precautions to minimize their adverse impact on your project. The most common problems are:

- Subs late.

  Let subs tell you when they'll be there.

  Call the sub the day before he is scheduled to arrive at the site.

Meet the sub at the site at a certain time.

Give incentives for early completion. Cash talks

■ Subs don't show at all.

Let them know up front that you expect prompt attendance and that you will spread the word around if they cancel with no warning.

Have other subs as backups.

■ Subs proceed slowly,

Give subs a bonus for getting work done on or before schedule. A cash bonus makes a great incentive.

■ Subs do work incorrectly or incompletely, or disagree on specific work to be done.

Ask them to repeat instructions back to you.

Have detailed specifications and drawings where needed.

Don't let too much work get done without checking on it. Visit the site daily.

Discuss detailed drawings with sub and make sure they understand it.

Have a site phone so that the sub can call you when questions arise.

Have detailed specifications and drawings where needed.

■ Dispute arises over amount paid or to be paid, especially when changes are involved.

Put payment and payment schedule in writing and have it signed by both parties.

All changes to the original plans should be written on a change order form and signed by both parties.

Use lien waivers to control your liability.

Pay with checks to insure proper records.

■ Materials not available when needed.

Order custom pieces early to insure that they are on site when needed.

Confirm delivery dates with suppliers and have them notify you of delays several days in advance.

Determine which suppliers deliver after hours and on weekends.

Get friendly with the delivery men and schedulers at your supply houses.

■ Wrong material or wrong quantity of material is delivered.

Use detailed descriptions or part numbers on all purchase order items.

Print legibly.

Don't order over the phone unless extremely urgent.

Have materials delivered early so they can be exchanged or corrected before they are actually needed.

Have your subs help you to determine actual quantities and materials.

Get to know your salespeople and work with the same person every time.

■ Bad weather.

Don't begin project during or just before the rainy season.

Get roof up as soon as possible.

Make the most of good weather.

Provide a drive area of #57 crusher run to avoid muddy areas.

Get a roll of poly (plastic sheet) to cover materials.

■ Theft and vandalism.

Visit site each day.

Call the local police department and have your site put on surveillance.

Don't always visit site at a strict routine time. Make unannounced visits - especially near quitting time.

Don't keep too much material on site. This is especially true for windows, doors, millwork, good plywood, and expensive tubs and appliances. Some builders even put a three-inch screw in the top corners of a stack of plywood.

Get Builders risk and theft insurance with a low deductible.

Install locks as soon as possible and use them.

RULES FOR SUCCESSFUL BUILDERS

■ PUT EVERYTHING IN WRITING - *not written, not said.*

■ BE PREPARED *before* breaking ground and don't get behind.

■ KEEP PROJECT BINDER current with all project details.

■ BE FAIR with money. Pay subs what was agreed, when agreed.

■ ALWAYS GET REFERENCES and check out three on each major sub. You must see their work.

■ DON'T try to do too much of the actual physical work yourself. Your time and skills are often better used managing the job.

■ SHOP CAREFULLY for the best material and labor prices. Don't hesitate to insist on builder discounts.

■ USE CASH to your maximum advantage.

■ RELY ON YOUR CONTRACTORS for advice in their field. But above all, use common sense.

■ DON'T ASSUME anything. Material is not on site unless you see it. Subs are not on site unless you see them there. Don't expect subs to know what you want unless it's in writing.

■ USE as few subs as possible without sacrificing cost or quality.

■ USE the best materials you can afford.

■ KEEP CHANGES TO A MINIMUM and budget funds for them carefully.

■ ESTIMATE on the high side and have a contingency fund.

■ KEEP money under control and track progress versus your original budget.

■ RELY on your local building code as a primary reference for specifications, inspection criteria, material guidelines and settling of disputes. Learn to use it comfortably.

■ STAY at least a week or two ahead of the game. Such a project horizon will allow you time to compensate for material delays or time needed to replace unreliable or unsatisfactory subs.

■ USE THE BEST MATERIALS you can afford.

■ LET your fingers do the walking. If it looks like rain, call the roofer. If something can't be delivered, advise your subs.

PROVERBS FOR BUILDERS

Below are a few sayings very appropriate to the building industry. As a builder, you are bound to encounter times when one or more of them apply.

MURPHY'S LAW. If anything can go wrong, it will.

IF there is a possibility of several things going wrong, the one that will cause the most damage will be the one to go wrong.

If you perceive that there are four possible ways in which a procedure can go wrong, circumvent it, then a fifth way will promptly develop.

It is impossible to make anything foolproof, because fools are so ingenious.

DAVE'S CONSTANT. Matter will be damaged in direct proportion to its value.

If you explain so clearly that nobody can misunderstand, somebody will.

Do not believe in miracles...rely on them.

LESTER'S FIRST LAW. Everything put together falls apart sooner or later.

LESTER'S SECOND LAW. No matter what goes wrong, it will probably look right.

If only one bid can be secured on any project, the price will be unreasonable.

KING'S LAW. No matter how long or how hard you shop for an item, after you've bought it, it will be on sale somewhere, for less.

Everything takes longer than you think.

THE GOLDEN RULE: He who has the gold makes the rules.

The Ninety-Ninety Rule of Project Schedules: The first twenty percent of the task takes ninety percent of the time and the last ten percent of the time takes the other ninety percent.

Once you have exhausted all possibilities and fail, there will be one solution, simple and obvious, that is highly visible to everyone else.

All delivery promises must be multiplied by a factor of 2.0.

HARRISON'S LAW: The most vital dimension of any plan or drawing stands the greatest chance of being omitted.

Commoner's Law: Nothing ever goes away. When things just can't get any worse, they will. Left to themselves, things tend to go from bad to worse.

Nothing is as easy as it looks.

WHY CONSTRUCTION PROJECTS FAIL

As a builder, it is up to you to make your project a success, but along with all the glory comes all the responsibility! Listed below are a number of common reasons why construction projects fail and/or encounter severe difficulties:

■ Costs are underestimated.

■ Lack of budget control. Excessive design changes are made late in the game.

■ Project manager not organized; letting things get out of control.

■ Lack of understanding of contracting.

■ Acts of vandalism, arson or natural catastrophes not adequately covered by insurance.

■ Lack of enthusiasm.

■ Workers paid before careful inspection was made of work and materials.

■ Legal complications due to inadequate compliance or careless handling of business transactions. Such as double invoice payment or payment for work not completed or undelivered, damaged or inferior goods.

■ Conflicts with spouse or partner.

BUILDING FOR YOURSELF VS. BUILDING FOR OTHERS

One of the first questions that any of your workers are going to ask you is whether you are building this home for yourself or to sell. The difference can be subtle, but more likely the difference will change the house dramatically in terms of workmanship and materials used.

Those who build for a living learn quickly that the only way to make a fair profit is to build into a home ONLY those features and the level of quality that is necessary to sell the house and nothing more. Quality materials such as hardwood floors, marble, custom millwork, brickwork, fancy lighting and fancy bath fixtures are used sparingly or not at all.

However, you may defeat at least part of the purpose of building your own home if you don't outspend the neighborhood builder in a few areas. Perhaps you want a skylight, nice brass faucets, fancy wallpaper or better grade appliances. This is your prerogative. You should add a few special touches that will make your house stand out and easier to sell.

WHEN YOU GET NERVOUS

Nervous because you've never done anything so ambitious?   Nervous because the carpenter hasn't returned your call yet? Or maybe you just can't put a finger on the source of your uneasiness. RELAX.

REMEMBER:

Fear is a natural response to the uncertain and the unknown. A goal of this book is to replace uncertainty with clear, concise knowledge - to remove those cloudy uncertainties. Some individuals maintain a certain level of fear as a natural defense mechanism - to always remain cautious and alert of the situation and possible problems.

Individuals with little or no formal education build residential dwellings every day... and they do it for a living!

A certain degree of fear is typical when doing anything for the first time. Chances are, after you have built your first home, you'll want to do it again!

Remember, your construction project is staffed by many knowledgeable, professional sub-contractors each specializing in their area. It is their job to do the work properly. Don't worry if you don't know everything they do. That is not a requirement of a builder. If the job is not done right you can refuse to pay until the items in question are fixed.

The construction approach presented in this book has been time - tested. Many homes have been built using material in this book. Using this book puts you way ahead of many individuals who may either try to brave it alone or oth-

erwise push ahead without adequate knowledge and/or preparation.

Assure yourself that you are doing everything you can to protect your project against the pitfalls described in "WHY CONSTRUCTION PROJECTS FAIL".

Assure yourself that you are using the STEPS, SPECIFICATIONS and INSPECTION checklists provided.

WHEN YOU GET DISCOURAGED

You're into the project and things aren't going along perfectly. Your framing crew made some mistakes, you've had four straight days of rain, your roofing sub won't return your call and the price just went up on storm windows. So what else is new? Getting discouraged?

Did you ever hear of any project proceeding from beginning to end without a hitch or two - Or twenty? Below are a few things to ponder when you wonder why you didn't just throw this book away and cut your losses short.

LOOK at the money you're saving. You probably have a good idea how much. Just look at that figure for a moment. That's the money you would have paid someone else to do the work you're doing right now. If someone else were performing your job, they'd run into the same kind of problems. When the interest costs are figured in, you will probably be saving one or more year's salary by contracting yourself. Aren't a few months of challenge worth it?

LOOK at all the problems and solutions you're encountering. One day when your home is completed, you'll look back at them and laugh. You'll tell of some of your experiences the rest of your life.

LOOK at your accomplishment! Your friends, associates and family look upon you with admiration. You can honestly say that you built your home. Only you will know how much physical work and skill you didn't put in).

REMEMBER - Lesser days are always followed by better ones. Hang in there!

2. Legal and Financial Information

Obtaining financing for your home will be one of the most critical and demanding phases of your building project. You must be bookkeeper, financier, and salesman. The lender will want evidence that you know what you are doing and that you can finish your project within budget.

THE CONSTRUCTION LOAN

Unless you are one of the few who build with cash, the construction loan is a necessity. The construction loan process will be a trying one, entitling you and a friend to a night out upon loan approval. It is typical for interest on a construction loan to be a few points higher than a permanent loan. This gives you further incentive to finish the project as quickly as possible. WARNING: When you apply for your loan, apply for all the money you will need - and then some. Don't plan on getting extra money later. You are headed for a disaster of mammoth proportions if you run out of money before your project is over. Your lender will foreclose on your project and you'll end up with next to nothing. This also means that once you've done your take-off and are into construction, don't make expensive changes and upgrades unless you can pay for them out-of-pocket.

THE FINANCIAL PACKAGE - APPLYING FOR YOUR LOAN

Make sure that your financial package is complete before visiting the lender. The package should be neat, informative, well organized, and comply with all of the lender's requirements. Your lender will want to be sure that you can keep good records. Your financial package should include the following:

- Personal financial statement.
- Resume or personal biography.
- Total cost estimate of the project.
- Description of Materials.
- Survey of the proposed building site showing location of dwelling
- Set of blueprints

THE PERSONAL FINANCIAL STATEMENT should include all your assets and liabilities. Make sure to include everything of value you own including personal property. Many people tend to underestimate the value of items like furniture and clothing. Your liabilities should include any loans, charge accounts and credit balances. Ask your lender for one of their own standard forms.

THE RESUME is designed to sell yourself to the lender. Start with a standard job resume but be sure to emphasize any construction interests or experience.

THE COST ESTIMATE will be very important as your construction loan will be based on it. Make sure to include a 7 to 10% cost overrun factor in your estimates as a cushion for unexpected expenses. Many lenders can provide you with a standard cost sheet for you to fill out. If not, use the one included in this book.

THE DESCRIPTION OF MATERIALS form will aid the lender's appraiser in estimating the appraised value of your property. Many lenders use the standard FHA form for this purpose, but check to see if your lender uses their own special form.

THE SURVEY should be obtained from the seller of the property or from the county tax office. If the property is very old, you may have to request that a new survey be done. Indicate on a copy of the survey the approximate location of the house.

A SET OF BLUEPRINTS complete with all intended changes should be submitted. Give your lender the set without the coffee stains.

START-UP MONEY

You will need to put up a lot of your own money before you can expect to receive your first draw. Your initial outlay will include:

- Lot cost to be paid before the construction begins.
- Closing costs (construction loan closing)
- Permits (building, water, sewer, etc.)
- Site preparation (clearing, grading, and excavation)

- Foundation

If you can't do this out of your pocket before going to the lender, you aren't ready to begin building.

ABOUT DRAWS

Unlike a car loan where you receive all your cash at once; construction loans are almost always handled on a draw basis. The lender will make you earn the money that you borrowed - bit by bit. Each time the lender sees that you have completed say, 5%, of the project, you are allowed a "draw" of 5% of the construction loan. This nuisance is, in fact, a way of making sure that money isn't wasted carelessly up front. Whenever you would like a loan draw, you must contact the lender and request one. The lender has an inspector that will visit the building site to see what work has been completed. The lender will normally have a fixed draw schedule which allows a certain amount of money to be available for each major phase of construction. Many lenders inspect on Thursdays and distribute draws on Fridays. Builders are usually allowed only about five draws free of charge. If you desire more draws than allowed per project, you are usually charged a nominal fee per draw.

Money is drawn after proof is shown that such money is deserved. Hence, you must go into the project with some start-up money of your own. The need for this "pocket money" can't be underestimated. It will be used to buy permits and licenses, dig your foundation and perhaps get your foundation built before money is drawn. It will also be used to cover unexpected expenses, such as price increases, last-minute additions you may want to make, and underestimates. Be aware that draw schedules are normally back ended; meaning that you will spend more than you can draw up front, while receiving a 5% draw for painting your mailbox near the end.

THE CONSTRUCTION PERMANENT LOAN

When possible, apply for a Construction Permanent Loan (also known as a convertible loan). This type of loan eliminates one of two closing costs related to separate loans for construction and permanent financing. This is done by putting both the construction loan and the permanent loan in your name. When the permanent loan is ready to close; the construction loan is "converted" to a permanent mortgage. Shop loans carefully. Time here is well spent. Many institutions now offer 15-year payoffs that only cost a little extra each month over a standard 30-year loan. Points, a lump-sum cost charged when you start a loan, is only a way for lenders to get a little bit more money out of you without raising the interest rate.

DEALING WITH THE LENDER

When going to the lender, be prepared:

■ Have your financial package ready.
■ Know exactly how much money you want to apply for.
■ Have all requested information typed.
■ Dress up (Preferably dark suit).
■ Show up at the scheduled time.
■ Don't mention how much pocket money you have on the side - the lender will want you to put it into the construction loan.
■ Don't talk too much. Answer questions only when asked and don't volunteer any information not asked for.
■ Have a list of your subs and backups.
■ Have all lender forms typed.

THE APPRAISAL

The lending institution will have its capital at risk. As such, they will want to be assured that the house you are planning to build is worth more than the construction loan. They will also test your skill at estimating costs.

The lender will appraise your home in one of two ways; the Market Approach or the Cost Approach. Most likely, the lender will appraise your home based on the Market Approach.

Market Approach - The method of appraisal bases the value of the home on "comparables" in the area. The appraiser will look for comparable houses to yours in the surrounding neighborhood that have similar features, style, square footage and property values. Therefore it is advantageous to build your house near a neighborhood with homes similar but higher in price than yours. Don't get too far fetched with your design. The appraiser may not be able to find suitable comparables and may have to guess at the value; usually to your disadvantage.

Cost Approach - This is either similar to the cost take-off or a determination of value by square footage.

After the value is determined, the lender will then loan you a fixed percentage of the appraised value; usually up to 70 or 80%, to build the house. Very seldom will the lender give you a loan based totally on what you claim your construction costs to be. However, you should easily be able to build a house for less than 75% of its appraised value.

THE LEGAL ISSUES

Real estate laws have evolved over thousands of years. Needless to say these issues are complex and are best left to the experts. Make sure to consult the appropriate authorities about any legal issue in order to avoid unpleasant surprises. The professionals you will need to deal with are attorneys, insurance agents, and financial lenders. The key legal issues affecting your building project are:

> Filing in the public record
> Free and clear title to the property
> Land survey
> Easements
> Covenants
> Zoning
> Variances
> Liens
> Sub contractor contracts
> Builder's Risk Insurance
> Building permits
> Workman's Compensation

FILING IN THE PUBLIC RECORD - Because of the importance and complexity of Real Estate law, all real estate transactions concerning the transfer of property, liens, and loans are normally filed in the public records at your county's local courthouse. This way you are announcing to the public what property you own. Anyone has access to these records and may research the history of your property to see who owned it before you and if there are any mortgages, judgments, or liens on the property. This is for your protection; because you can do the same research on any property you intend to buy. Go to your local courthouse and familiarize yourself with the records. Ask one of the title lawyers in the room (there are always a few attorneys there) to help you learn how to search the records. This will give you an idea of the process of real estate transactions.

FREE AND CLEAR TITLE concerns your bundle of rights to a piece of property, whether it is land or improved property. Before closing on a sale of property, your lawyer will do a "title search" of the property in the public record to determine its ownership history. Most real estate records are kept at the county courthouse and are accessible to anyone. If the lawyer finds someone with an "interest" in the property or a confusion over rights, this will create a "cloud on the title" which must be removed before title can transfer. A title search is required by all lenders and is usually done for a standard fee. *NEVER BUY A PIECE OF PROPERTY WITHOUT DOING A TITLE SEARCH!* Many so called "bargains" are properties with title problems that the present owner is hoping to "pawn off" on some unsuspecting buyer. Don't be that buyer! You should also get title insurance, protecting your claim to the land.

A LAND SURVEY is another required item that must be done before title can be transferred. This surveyor can usually be arranged by your lawyer or financial lender. The survey will describe the exact location of your land in a written legal description and scale drawing of the property called a "plat map". These descriptions will be attached to the deed and filed in the public record.

EASEMENTS on the property will normally be discovered during the survey and title search. Easements give rights of traverse on your property to others such as local governments (sidewalk easement), utility companies (sewer or power line easements), or individuals. Make sure to check the location of your construction in relation to the easements to avoid any conflicts.

COVENANTS are building restrictions or requirements placed on construction in certain subdivisions such as minimum square footage of the building or the type and style of construction. These restrictions are usually placed on the property by the developer of the subdivision in order to protect the value of the homes in the subdivision. Ask your lawyer to check for any building restrictions in the public record before buying a building lot. You may not be able to build the house of your dreams if the design is not allowed in the neighborhood! Watch out for any restriction stating that the home can only be built by a specific set of builders or in a specific time frame.

ZONING RESTRICTIONS, like covenants, control use of land in an area and the type of structures built. These restrictions are placed on property by the local government planning board in order to protect the value of land in the area. Check with your local planning office if you are unsure of the zoning of your property. A zoning map of the county can usually be purchased for a nominal fee and can be very helpful in searching for property. Zoning laws also define the area of the lot that the structure can actually occupy. The builder is required to place the structure a certain minimum distance from neighboring homes. This requirement is called the minimum setback. In many planned sub-divisions, these boundaries are closely adhered to in order to make the homes in the area appear consistent and planned. You may also have requirements for a backyard of a certain depth and a minimum side yard of twenty feet. It is important to find out where your baselines are before you break ground. If you build outside a baseline and are caught, you may be in for a hefty fine.

VARIANCES allow you to "vary" from standard state and local zoning ordinances within prescribed, approved limits. For example, if you want to build your home thirty feet from the front curb and the local zoning ordinance requires a forty foot set-back, you must apply for a ten-foot set-back variance. Unless your request adversely affects neighborhood appearance or safety, your variance will normally get approval. To obtain a variance, you must normally pay an application fee and go before a panel of county zoning officials at a variance hearing where you or your representative describe or illustrate your intentions and the reasons for them. Normally a notice of your variance request will be posted on your lot so that the public is aware of your intent to get a variance. They have a right to appear at the variance hearing to either support or fight against your variance.

LIENS can become extremely important to a home builder because of the effect it can have on the construction project and the subsequent closing of the property. A lien places an interest in a property that will create a cloud on the title, preventing the close of the sale until resolved. For example, if the plumbing contractor does not feel that he was fully paid for his work, he may file a lien in the public record on the property involved. This lien can stay attached to the property until the disagreement is resolved. Many states do not require that the owner be notified of the lien; but you will know about it at closing! The lien will prevent sale or mortgaging of the property until the disagreement is resolved.

To avoid the pitfalls of liens to potential homebuyers, many states now require contractors to sign an affidavit guaranteeing to the purchaser that the contractor has paid all his bills due suppliers and sub contractors. If any parties then file a lien, the contractor can become criminally or legally liable not only to the lien holder, but also to the purchaser! Builders or subs can be thrown in jail in many states for this offense!

If you are building for yourself or others, you can protect yourself from liens by requiring your suppliers and sub contractors to sign a sub contractors affidavit. This agreement works like the contractor's affidavit. The sub contractor agrees that all bills have been paid and that the contractor has no legal claim against your property. This will prevent the sub contractor from filing any liens on the property. Make everyone you deal with sign this agreement for your own protection!

SUB CONTRACTOR CONTRACTS like affidavits, protect you from surprises. Get everything important in writing no matter how good your sub's memory or integrity. Subs may have a contract amnesia if a problem comes up. Make sure to include the specifications for the work done as well as a schedule of how payment will be made. It is wise to use a statement such as "Final payment will be made when work is satisfactorily completed".

BUILDERS RISK INSURANCE is another "ounce of prevention" necessary before starting your project. This insurance will protect against liability if someone should be injured on the building site. Theft of building materials is sometimes covered also. Risk insurance may only cover theft after the dwelling can be locked. Be aware of very high deductibles.

BUILDING PERMITS must be obtained from your County Building Inspector's office before construction can begin. This will notify the city or county of your project so that the building inspector can schedule the required inspection of the property during construction. This permit must be displayed in a prominent location at the site. Check your local county for details. Normally, building permits cost a certain amount per $1000 of construction value or per square foot. Obviously this value is somewhat intangible. You may have to accept the authority's estimated value.

WORKMANS COMPENSATION is required in every state to provide workers with hospitalization insurance for job related injuries. This accident insurance must be taken out by any person or company with hired employees. Most states also have special statutes that require builders, even when using subcontractors, to have this insurance protection for anyone working on the building site. At the end of each year, the insurance company does an "audit" to determine the amount of insurance to be paid. Each building trade is assessed an insurance rate depending on the relative risk involved in the trade. For instance, roofing trades will be charged a higher rate than trim carpenters. Since you are providing insurance to cover these workers on your site, make sure to deduct the amount of the insurance from the price paid to the subcontractor.

Many of the subs that work for you will have their own workmans compensation policies since they may have employees working for them. If a sub has a policy, make sure to obtain the policy number and expiration date. Most insurance companies allow you a credit on your own rate if the subcontractor has his own policy. This credit will be applied at the insurance audit. Contact your local insurance agent for more information about workmans compensation in your state.

Don't underestimate the importance of this policy. You are liable for any work related injuries that occur on your property. Without this policy, a negligence lawsuit could take your life savings. Look for any situations that can void Workmans Compensation coverage. For example, in some states, drywall subs experiencing accidents while using drywall stilts are not covered. Better safe than sorry!

<u>Notes</u>

3. Selecting Your Lot

The lot you purchase will be a part of the home. It's a location and setting you had better be comfortable with. There's an old saying about the three most important factors when buying a lot; location, location and location. While this stretches the truth, it does make an important point.

CONSIDERATIONS
If you have not purchased your lot yet, you may wish to ask yourself the following questions:

Would you prefer to build in a planned development (sub-division), where home values will be more stable and controlled, or on a one of a kind lot? If you are purchasing a lot in a subdivision, make sure that the building covenants work for you; not against you. Do the covenants allow you to build the home you want? When you want? Do the covenants allow you to sub-contract the home yourself? Do the covenants protect against lower-priced homes which could affect your resale value? Don't build in an area where homes are more than 15 or 20% lower than your planned home.

Would you prefer a large lot, which may require lots of time and money in maintenance, or a smaller one that may cost less and involve less maintenance?

Is the lot and surrounding area suitable for the type home you are planning to build? Consider this from both a physical and esthetic point of view.

Will the lot support a basement, slab or crawl space if you plan to have one? A better location is almost always worth spending a little more money.

ADVICE
Don't attempt to save money by buying a "less desirable" lot. Curb appeal can affect the value of your home by 25% or more. Saving $2,000 in lot costs could cost you $10,000 in home resale value. (Don't step over a dollar to pick up a dime!). Don't overbuild for the area; build in an area of comparable value. Make sure that the lot and the neighboring homes are WORTHY of the home you plan to build.

Obtain a title search of the property before purchasing to ensure free and clear title to the land. Make sure that there is public sewer access on the lot if the seller says it is available. Don't take his word for it.

It is wise to first determine the style and size home you plan to build before looking at lots. It would be a shame to find out that the home you want to build won't even fit on the lot after considering baselines. Also, you may want to build near homes of comparable value. To do this, you need to know what size home to build. This is the subject of the next chapter.

LOT EVALUATION

Review the following checklists of strengths and weaknesses although many of them may already be obvious to you. Each weakness provides grounds for negotiating a lower offering price. Once you own the land, each weakness is your problem.

STRENGTHS: CHECKLIST

_____ On sewer line. Check height of sewer line. Your lowest drain pipe should be slightly higher than the sewer line.

_____ On cul-de-sac or in other low traffic area.

_____ Well-shaped square, rectangular or otherwise highly usable.

_____ Trees and woods. Hardwoods and some larger trees.

_____ Good firm soil base. Important for a solid foundation.

_____ Not in flood plain.

_____ Safe, quiet atmosphere.

_____ Attractive surroundings such as comparable or finer homes and attractive landscaping.

_____ Level or gradual slope up from road.

_____ Good drainage.

_____ Lot located in stable, respected neighborhood.

_____ Just outside of city limits. May pay less local taxes.

_____ Easy access to major roadways, highways and thoroughfares.

_____ Convenient to good schools, shopping centers, parks, swim/tennis, firehouse and other desirable amenities.

_____ Situated in an area of active growth.

_____ Provides attractive view from prospective dwelling site.

_____ Fully usable.

_____ Seller will finance partially or in full.

_____ In area zoned exclusively for single family dwellings.

_____ Underground utilities. Phone, electrical, cable TV.

_____ Natural features in areas that will remain undisturbed.

_____ Swim, tennis, golf and other amenities.

WEAKNESSES: CHECKLIST

_____ No sewer, must install septic tank. Must pass a percolation test.

_____ On or very near major thoroughfare. Yellow or white lines down the center of the road. Speed limit over 35 mph.

_____ Odd-shaped (too narrow or too shallow). Either will cause limited front or side yards.

_____ No trees or barren.

_____ Rocky. May have large rocks to excavate or dynamite.

_____ Sandy soil. May require special foundation work. May not support a large foundation. Excavation difficult.

_____ In flood plain. Normally involves expensive flood insurance and difficult resale.

_____ Near airport, railroad tracks, landfill, exposed electric power facilities, swamps, cliffs or other hazardous area.

_____ Near large power easements, commercial properties or radio towers.

_____ Slopes down from road and/or steep yard. Hard to mow. Difficult access by car when iced over. Water drains toward home. May require excessive or unusual excavation and/or fill dirt.

_____ Flat yard. Less than 2% slope may be difficult to drain.

_____ Water collects in spots. May require expensive drainage landscaping and/or fill dirt. May result in a wet/damp basement.

_____ Lot located in or near unstable/declining neighborhood.

_____ Just inside city limits. May have to pay more local tax.

_____ Relatively unaccessible by major roadways.

_____ Isolated from desirable facilities such as shopping centers, swim/tennis, parks, etc....

_____ Lot provides unattractive views. Either from prospective dwelling site or from road.

_____ Creek, gully or deep valley running near center of lot.

_____ Area zoned commercial, duplex, quadruplex or other high-density, multi-family dwelling.

_____ Area populated by less expensive homes.

_____ Utilities above ground or not yet established. Unsightly electrical wires.

FINDING A LOT

Finding the right lot may involve approaching lots from any of the sources below and others as well:

■ Local real estate agents and/or multiple listing services

■ For Sale ads in newspapers

■ For Sale signs on land

■ Legal plats in county courthouses or other local government offices.

PURCHASING THE LOT

Review the considerations and the checklists on the previous pages. Check for problems and weaknesses. REMEMBER: "Once you buy it, it's YOUR problem!"

Request the seller to provide a legal description of the lot, prepared by a registered surveyor, showing all legal easements and baselines (you may not be able to build just anywhere on the lot). Negotiate legal and other closing costs to be paid by or split with the seller. If you must pay all or part of the legal fees, INSIST on appointing an attorney of YOUR choice.

CASH TALKS - especially when purchasing from an individual rather than an organization. Look for owner financing.

Insist on and pay for a title search and title insurance. Title insurance is important if there is ever a dispute over land ownership. Litigation could cost you lots of time and money. When applying a construction loan at a local lending institution, you must have "free and clear title" to the land. This insures that no dispute over title will interfere with the completion of a successful construction project. All financial institutions insist on you having free and clear title before negotiating a loan.

Notes

4. Choosing Your House Plans

HOW MUCH HOME CAN YOU AFFORD TO BUILD?

The standard rule of thumb used by most banks is that your principal and interest (P&I) payments should not exceed 25% of your Gross annual income. With interest rates fluctuating, many banks will now allow up to 35% of your income. To find the total amount of home you can afford, use the following equation. Use a mortgage payment table (available at most banks) to figure your monthly P&I.

EXAMPLE:
John earns $37,000 per year and has just bought a subdivision lot for $12,000. How much can he spend on house construction and still meet the bank's loan requirements? The bank is willing to finance 80% of the value of the house. The interest rate is 12%.

Monthly Gross income/4 (25%) = Monthly mortgage payments

(37,000/12) 3083.33/4   = 770.83

Look up the above payment in a mortgage chart to determine the amount of the mortgage.

Mortgage amount/percentage financing = total value of house

75,000/ .80   = 93,750

Many times you will be able to secure 100% financing on your own construction once you have free and clear title to the land. In this example however, we will use 80%.

Now take the total value of the house and subtract the cost of the land, construction loan cost, and closing costs to determine the total amount of money you can afford to spend on the construction of your home. As a rule of thumb, the minimum cost of your home, excluding land, will be roughly $30.00 per square foot of heated space (excludes garage). For expensive brickwork, appliances, cabinets, etc... costs can exceed $50.00 per square foot. You should check with local building associations to find out the typical cost per square foot in your area.

Total cost of home	$93,750
Land Cost	-$12,000
Construction interest	-$ 2,500
Closing costs	-$ 1,800

Construction cost	$77,450

This amount should be the upper limit that you are willing to spend to secure a home of your own in this example. Next, determine what it is costing builders to build in your area. One way of doing this is to ask a few builders what they would sell their homes for direct. Deduct about 15 to 25 percent profit for them and then divide by the square footage. This will give you the cost per square foot. Another way is to ask experienced salesmen at a large building supply outlet. They always talk to builders and know what's going on. Remember, the cost per square foot does not include the cost of the lot.

Back to our example; the total cost of the home less the cost of the lot is $81,750. Assuming that the cost per square foot is $31.25, you could build a 2515 square foot home. This is quite a bit of house at your cost! Your actual cost per square foot will obviously depend on the quality of your materials and the extras that you include. Only after you have done your cost take-off will you know for sure what your house will cost, but this figure will give you an idea of the size house you can afford. Keep your square footage figure in mind when deciding on a house design.

BUILDING VS. REMODELING

If you're looking to save money, you might be considering remodeling an old home instead of building a new one. This can be especially attractive if a low interest loan can be assumed on the property. Examine this option carefully. Remodeling can have many pitfalls! Virtually all information in this manual, except for buying land, is useful to remodelers.

ADVANTAGES OF REMODELING

Low interest rates can sometimes be assumed on older homes.

There is a high potential for finding good bargains.

The property can usually be occupied sooner than new construction.

Tax laws provide numerous incentives for remodeling, especially in areas designated as historical districts.

Additional money can be saved by doing much of the work yourself.

DISADVANTAGES OF REMODELING

Many sub contractors are hesitant to do remodeling because of the hassles involved. Good remodeling contractors are hard to find and many require payment by the hour.

The true cost of remodeling may be hard to determine before starting the project due to hidden factors (rotten wood, air infiltration, foundation flaws, poor wiring, poor plumbing, etc.).

The improvements on the house may upgrade it to a value above the surrounding neighborhood. The value of neighboring homes may hold down the resale value of your home. Consider though, that this can also work in your favor if you buy the smallest home in the area.

Appraisers have a hard time determining the final value of a home before the remodeling is done.

Future utility costs are usually higher due to less insulation which can be difficult or impossible to improve on.

Remodeling is more suited to someone who plans to do much of the work himself.

If you plan to remodel an existing home, get several bids from remodeling sub contractors and have the house inspected closely before making your offer. Then make your sales contract contingent on a good appraisal, approval of the remodeling loan, and the discovery of no hidden defects.

ESTIMATING REMODELING COSTS

If you contract a "remodeling specialist", a company with many skills, you are paying retail costs. They will be their own general contractor on your project. Often, their bids will range from two to three times the sum of material and labor costs. By doing your own contracting, your cost will normally be a fraction of this.

WHAT TYPE OF HOME SHOULD YOU BUILD?

CHOOSE A HOME BASED ON YOUR LIFESTYLE

Up front, some people choose their home based on what they believe other people like. Of course, this defeats the purpose of "having your own castle". Make sure that you consider what you want based on the way you live.

Before deciding on a plan, make a list of those features you find most valuable to your own style of living. This will be your set of criteria for evaluating plans. Do you like a formal setting or a more rustic environment? Is a formal dining room or living room important to you? What are your family considerations? Listed below are some suggestions to keep in mind when making your planning decision.

- Eliminate as many hallways and walls as possible. Hallways waste building materials and add to heated space while making a house seem smaller.

- Combining living and dining rooms into one room will make a smaller home seem more spacious and will cut down on building materials needed for separating walls.

- Place the kitchen near the garage or driveway to provide easy access when bringing in groceries.

- Two story and split level homes are generally less expensive to build than ranch style houses of the same square footage because there is less roof and foundation to build.

■ Make sure the inside of the kitchen cannot be seen while entering or sitting in the living area.

■ Use closets as sound buffers between adjoining rooms.

■ Avoid house plans that have lots of different sizes of windows. Six or seven different sizes should suffice. During purchasing, delivery and installation, many frustrating mismatches can occur.

ENERGY CONSIDERATIONS

The energy efficiency of a home can be greatly improved with a little advance planning. Listed below are a few energy considerations:

■ A window's insulating ability is much lower than that of a typical wall. Reducing the amount of window coverage will increase insulating efficiency. This can be done by reducing the number of windows (which also saves money) or the size of each window. Wooden windows, although more expensive tend to insulate better then metal ones. Choose double or triple pane windows or use storm windows for the best insulation possible.

■ Make sure insulation has a vapor barrier - either foil backed insulation or a sheet of plastic placed between the interior drywall and insulation. The vapor barrier prevents condensation from building up in the wall. Plastic sheeting over unfaced insulation is generally cheaper and more energy efficient.

■ Framing your exterior walls with 2 x 6's 24" on center will allow 6" batts of insulation to be placed in exterior walls, but will cost more in lumber costs. Also, framing in doors and windows can be a problem with 6" walls. An easier way to achieve a higher insulation value is to use thinner walls but more efficient exterior sheathing such as foil backed urethane.

■ Orient your building site so that the exterior walls with the smallest window area face north. Southern exposure to sunlight will allow more solar heat gain during the day.

■ Planting hardwoods near the house will provide shade in the summer, but will allow the sun to shine through in the winter when the leaves fall. Don't plant trees too close to foundations, driveways, patios or walkways or you will eventually face the risk of cracks.

■ The two most efficient heating systems are gas heat and heat pumps. Gas heaters are cheaper to install and are simpler to maintain; however, heat pumps use less energy to operate. As gas prices are deregulated, heat pumps will become an even greater bargain. Heat pumps lose their efficiency in any climate where winter temperatures stay below 30 degrees. In this type of climate, gas heat may be a better bargain.

■ The importance of caulking air leaks cannot be over emphasized. Air infiltration can contribute up to 40% of a home's heat loss. Buy a can of urethane spray insulation and caulk around doors, windows, electrical outlets, switch plates, and pipes during construction.

SAVING ON FOUNDATIONS

Choosing the proper foundation can save money. The foundations listed below are the four major foundation types. They are listed in order from least expensive to most expensive. See the section on foundations for further information.

ALL WOOD FOUNDATION

An all wood foundation can be installed at any time of the year and saves labor costs because only one sub contractor (carpenter) is used. Acceptable for crawl spaces and full basements.

MONOLITHIC SLAB

A monolithic slab is the least expensive concrete foundation and a good choice for quality and savings. Work can be completed quickly for additional savings. Care must be taken to position plumbing accurately as it will be cast in the concrete.

CINDER BLOCK

Cinder block is somewhat cheaper than concrete but requires the use of several different subcontractors. This causes a longer completion period and more complications. Not as strong as concrete and more vulnerable to leaks.

POURED CONCRETE

Poured concrete is the best choice for full basements due to high strength and resistance to leakage. It is the most expensive type, but time and money can be saved because only one sub contractor is used.

COORDINATING SUB CONTRACTORS

The task of timing and coordinating subs can be difficult but can yield savings by reducing construction time. In some instances, promoting cooperation between subs can make their jobs easier and give you a reason to bargain for a lower price. One example would be coordinating the painter and trim carpenter. If trim and walls are to be of a different color, then have the painter put on his primer coats before trim and doors are installed. Primer coats can also be put on trim and doors before they are set. This will save the painter time and effort and will speed up the construction process.

HOUSE PLANS

HOW MANY COPIES? Regardless of how you arrived at a design, you will need plenty of copies of your plan. Plans refer to a standard set of blueprints showing basic elevations of the home and a layout of each floor. Special detail plans will be covered later in this section. The following is a representative list of the copies you may likely need:

Personal copies	2
Bank copy	1
County inspector	1
HVAC	1
Plumbing	1
Electrical	1
Framing	1
	--
	8

Sound crazy? Copies are cheap. Normally $10 after the first one. Don't scrimp here. Get what you need. It's always nice to have that extra clean copy when your originals become ragged from use. You can not directly make a blueprint from a blueprint.

GENERAL ADVICE

When building your first home, consider using a standard plan. Preferably a popular one designed by a reputable architect or plan company. Plans drawn by architects are generally engineered better but are more expensive. If you are staying with a simple standard layout, home planners and draftsmen's plans are usually fine and much cheaper. Go to the architect if you plan any special layouts such as special foundations or vaulted ceilings. Their expertise will be well worth it. Always use plans with a standard scale such as 1/4' to 1'.

If you build again, consider using the same plan again. You'll learn from your mistakes and probably do an even better job. You will also know exactly how much materials are needed. Your costs will be more accurate.

Critically review the kitchen and kitchen cabinet layout. Cabinet layouts are normally there just to meet minimum requirements. You can probably do better. The kitchen design itself may not be the best either.

For sources of houseplans, check with builder plan services, architects, draftsmen and new home magazines. Architects and draftsman usually charge a certain fee per square foot of house.

For an additional charge, your plan supplier may be able to provide you with a material takeoff. But DO NOT assume it to be accurate.

If you change interior partitions, be careful of any impact on load-bearing walls. Consult an architect if you are in doubt. If you want to finish the basement, opt for a nine foot basement ceiling to allow for a dropped ceiling to hide heating and ducts.

PLAN CONTENTS

How do you know if your plans are complete? At a minimum, a set of plans should include the following diagrams:

- Front elevation

- Side elevations (one for each side of home)

- Rear elevation

- Top view of each floor (floor plan)

- Detailed layout of kitchen cabinets

- Detailed foundation, footings and framing specifications

SPECIAL DETAIL PLANS

Although you have a set of blueprints, they may not include the following:

- Mechanical layout (HVAC and ductwork)

- Electrical layout

- Plumbing layout

These are normally done by the respective subs based in part on your specific lifestyle, tastes and ideas.

Your standard set of blueprints may also include:

- Detail Cabinetry Layout

- Trim Detail Layout

- Detail Tile Plan

DETAIL CABINET LAYOUT - This should include a final layout of each specific cabinet you plan to have in the kitchen, including style, finish and placement. All cabinetry in bath areas such as vanities must also be spelled out.

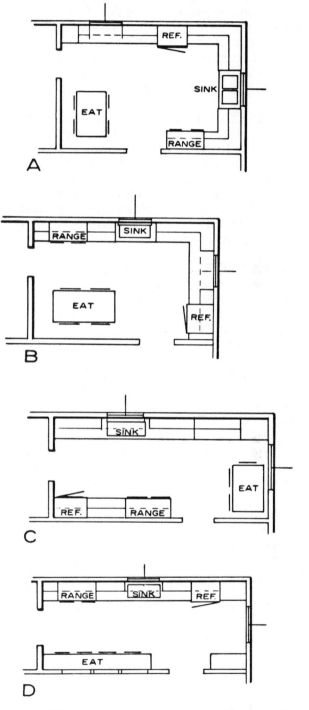

Figure 1. - Kitchen layouts: A, U-type; B, L-type; C, "parallel wall" type; D, sidewall type.

TRIM DETAIL LAYOUT - This should include a room-by-room schedule of what trim is used where. Base, chair rail, crown and other decorative molding as well as paneling should be described in detail. Specific material finishes should also be covered. Trim detail layouts are not usually included with most standard plans.

DETAIL TILE PLAN - This should include diagrams showing areas to be covered with what types of tile. All tiled shower stalls, bathtub areas, tiled walls and floors should be covered. Also, all fixtures such as mirrors, soap and grabs, safety rails, towel racks, tissue roll holders, toothbrush and tumbler holders, medicine cabinets and related fixtures. Grout types and colors also need to be determined.

WINDOW SCHEDULE - Itemized list of all window types and dimensions.

CHANGES

After you have selected your plans, consider special changes such as zoned Heating/AC, passive solar, special wall partitions, window modifications, special cabinetry, etc... Your original drawings should provide the basis for any details, refinements or other changes you wish to make. When you decide to change a plan, mark the change on the plan with a red permanent marker. (This will keep the change from smearing if the plans get wet.) Make sure when making changes on plans that you mark ALL the sets of plans to eliminate costly problems later. Many lenders require that houses be built exactly as the plans indicate.

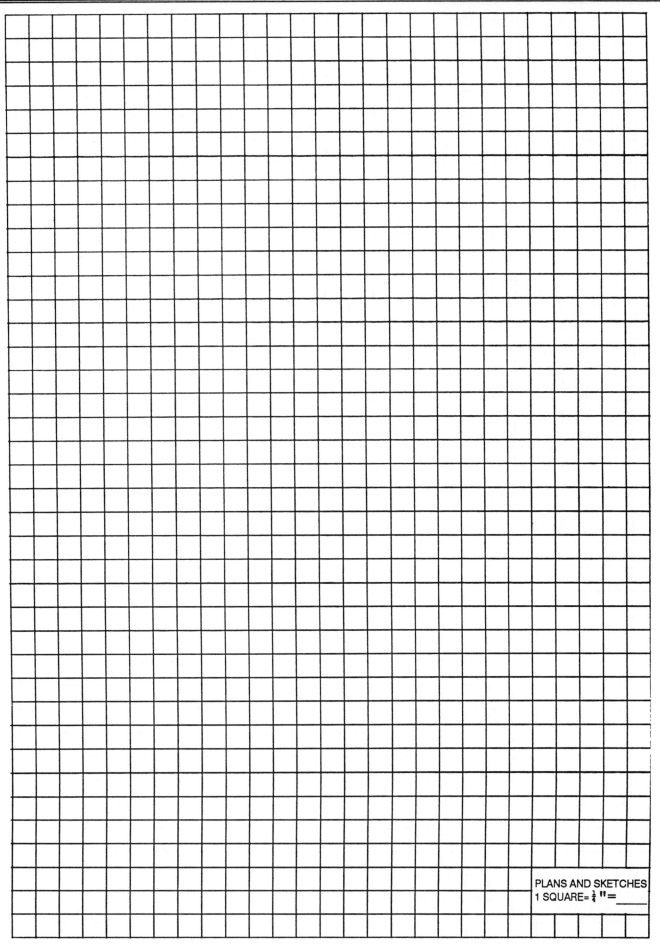

PLANS AND SKETCHES
1 SQUARE= ¾ "= _____

PLANS AND SKETCHES
1 SQUARE= ¼ "= _____

5. Dealing with Material Suppliers

DEALING WITH LARGER COMPANIES offers advantages due to potential savings passed on as a result of volume purchasing. Larger companies may offer wider selections than their smaller counterparts. Larger companies have a larger reputation to protect. Hence your satisfaction is of great importance.

DEAL DIRECT WITH MANUFACTURERS when possible. Companies that specialize in windows, doors, carpet, bricks or other such items may provide you with merchandise at close to wholesale prices. Be prepared to pick many of these purchases up yourself since delivery may be costly or unavailable.

LIMIT THE NUMBER OF VENDORS YOU DEAL WITH in order to reduce confusion, building accounts and persons you must deal with. Dealing with a few vendors may mean special discounts due to larger volume purchasing.

OPEN BUILDER ACCOUNTS at your suppliers. You are a builder!...And builders get discounts. Also, builders don't deal with retail sales personnel. Most large material supply houses have special contract sales personnel in the back of the store. Builder accounts often imply special payment schedules. For example, in one case you may pay the open balance on the first of each month. In this case it is to your advantage to purchase your materials early in the month to gain a full month's use of the money. The time value of money is on your side in this case. Paying accounts within a specified period is a sure fire way to earn early payment discounts. Keep track of them and earn them. That's money in your pocket.

INSPECT MATERIAL as it arrives on site. If something is damaged or just doesn't look right, insist on having it returned for exchange or credit. This is normal and expected: you're paying good money and you should expect good quality material in good condition. Indicate returned items on the bill of lading as proof.

GET SEVERAL BIDS on all expensive items or whenever you feel you could do better. The old saying "Only one thing has one price: a postage stamp", is alive and well in the building material business. Shop around and chances are, you'll find a better price. Check with the local builders in your area and ask them where they buy their materials.

DETERMINE WHAT IS RETURNABLE In the normal course of construction, you will more than likely order too many of certain items accidentally or on purpose to avoid delays due to shortages. Find out what overages can be returned to suppliers. Ask for written return policies. Certain cardboard cartons, seals, wrapping or steel bands may need to be intact for an item to be returnable. There may be minimum returnable quantities for such items as bricks in full skids (1000 units). Restocking charges may also be incurred when returning materials.

MANAGE YOUR PURCHASES WITH PURCHASE ORDERS. Explain to your suppliers up front that you plan to control purchases with "P.O.s" and that you will not pay any invoice without a purchase order number on it. Purchase orders help to ensure that you receive what you ordered and no more. Make sure to use two part forms. This makes it impossible for someone to create or alter a form. Pre numbered purchase orders are available at most local office supply stores.

MINIMIZE INVENTORY and schedule your material deliveries with the project schedule as best you can. This is easier said than done. Minimizing the materials on site minimizes your investment and exposure to theft or damage. Most material suppliers will be glad to store purchased items for you and deliver them when ready for use. Fragile materials such as drywall and windows are good examples of items not to have laying around too long. Try to be on site when material is delivered. Write special delivery instructions on purchase orders. You're the customer: you make the rules.

HAVE MATERIALS DROPPED CLOSE TO WHERE THEY WILL BE USED in order to save time. Bricks, loads of sand and gravel are good examples. Require delivery men to move plywood, drywall and other heavy items up to the second floor if that is where they will be used. Make sure to write your delivery instructions and directions to the site on the purchase order.

KEEP MATERIALS PROTECTED from weather by covering them with plastic or keeping them inside whenever possible. Lumber, especially plywood, is susceptible to water damage. Lumber should be covered with heavy duty plastic and should be ordered shortly before it will be used. Many suppliers will allow you to purchase materials and keep them in their warehouses until you need them. Take advantage of this. Rocks, brick or scrap wood should be used to hold the plastic down on windy days.

PROTECT MATERIALS FROM THEFT means not having material suppliers deliver until shortly before the material is needed. With stacks of plywood, some builders put a long wood screw through the top corners of the top sheet. If someone wants a sheet, they'll have to haul off four at a time. This usually stops kids. Make sure to lock up all movable items in a storeroom or basement. You will want to install the locks on your house as soon as possible to protect interior items from theft. Instead of giving subs keys, unlock in the morning and lock in the evening.

KEEP TRACK OF SALES TAX During the construction of your home, you could be spending literally tens of thousands of dollars on materials. If your state has a sales tax, this will likely amount to at least hundreds of dollars. This is can be a sizable deduction at the end of the year. To do this, you must be able to substantiate your deduction with all related documents. Make a habit of logging all of your sales tax and keeping all receipts.

GET DISCOUNTS from early payments. Many suppliers give a percentage discount if full payment is made by the end of the month, the beginning of the month or within 30 days of purchase. Make sure you don't lose these discounts. Take advantage of them because they can really add up. Deal with lumber yards and other suppliers you will deal with heavily where discounts are possible.

USE FLOAT as much as possible. Many suppliers will invoice you at a certain period each month and will only charge you a finance charge thirty days after that. Hence, if you pickup materials at the beginning of the month and they invoice you for it at the end of the month, you have had thirty days' free use of their money.

6. Dealing with Subcontractors and Labor

Labor generally refers to employees that are working for you and have employment tax and withholding taken from their pay. If you use employees, get a good bookkeeping service to figure withholding taxes for you. The subcontractors you use are in business for themselves and withhold their own taxes. Your subs may consist of workers that supply labor only such as framers and masons or businesses that provide materials and labor such as heating and air contractors. Use contractors whenever possible to avoid the paperwork headaches that go along with payroll accounting. Using contract labor will also save you money because you are not paying part of their withholding taxes. The IRS may require you to file a "1099" form on your subs. This form shows how much you have paid your subs so that the IRS can be sure that they are paying their share of taxes. Check with a Certified Public Accountant to obtain the proper forms.

LOCATING WORKERS
Finding good subs and labor is an art. The following is a sure fire list of sources for locating subs and getting you on the right track of determining which is the best combination of skill, honesty and price:

- JOB SITES - (this is excellent because you can see their work). As your building site shows progress, it will act as a billboard, attracting all sorts of subs looking for work.

- REFERENCES FROM BUILDERS AND OTHER SUB-CONTRACTORS

- THE YELLOW PAGES Look under specific titles. Do this only when you are in a pinch because these folks always seem to charge a bit more.

- CLASSIFIED ADS - here you can normally bargain more than when you meet them at a job site

- ON THE ROAD - look for phone numbers on trucks - busy subs are normally good subs

- MATERIAL SUPPLY HOUSES keep lists of approved subs. Many of them keep a bulletin board where subs leave their business cards.

REFERENCES are essential in the home-building business. Nobody should earn a good reputation without showing good work and satisfied customers. Insist on at least three references; one of them being their last job. If their last reference is four months old, it may be their last good reference. Some feel that companies which use a person's name as the title, such as "A. J. SMITHS's ROOFING" will tend to be a "safe" choice over "XYZ ROOFING" under the premise that a man's name follows his reputation. But this obviously cannot be taken at face value. Contact references and inspect the work. You may even wish to ask the customer what he paid for the job, although this may sometimes be awkward. A good local standing is essential.

BACKGROUND INFORMATION
Acquiring reference information should be a rule. It is not a bad idea to have your "candidate" sub-contractors fill out a brief form in order for you to find out a little about them. By asking the right questions, you can get a fairly good idea of what type of character they are.

ACCESSIBILITY
It is important to be able to reach your subs. Try not to hire subs that live too far away. If they need to make a quick visit to fix something, if may not be worth their time. Your subs must have a home phone where they can be reached.

PAYING SUBS
When you have your checkbook out to pay your subs, remember LESTER'S LAW - "The quality of a man's work is directly proportional to the size of his investment." If you want your subs to defy human nature and do the best job possible, then give them motivation. *DO NOT PAY ANY SUB-CONTRACTOR UNTIL YOU ARE COMPLETELY SATISFIED WITH HIS WORK AND MATERIALS.* Once you have paid him, you have killed all incentive for him to perform for you. After you have paid him, do not expect to see him again in your life. Subs make money by getting lots of jobs done; not by taking lots of time on one job. Make each of your subs earn their pay. Below is a suggested schedule for payment:

■ 45% after rough-in.

■ 45% after finish work complete and inspected.

■ 10% retainage held for about two weeks after finish work. (this is to protect yourself if something is detected a bit later).

Let them know your pay schedule in writing before you seal the deal.

PAYING CASH
Paying with cash gets results. Telling a sub that "you will pay him in cash upon satisfactory completion of his job" should be ample incentive for him to get the job done - and may motivate him to do a good job to boot! Make sure when obtaining cash for work that you keep documentation of the work done for the IRS! Write a check in the sub's name and have him cosign it. Then cash the check. This way you will have an "audit trail" to prove that you paid the sub for work done.

BE FAIR
Be fair with money. Pay what you promise. Be discretionary with payments. Pay your subs and labor as work is completed. Not everyone can go a week or two without a little money. It is customary to pay major subs, such as plumbers, HVAC subs and electricians 40% after rough-in and the remaining 60% after final inspection and approval. This keeps the "pot sweet". If you pay subs too much too early their incentive to return may diminish. If he doesn't return you'll have to pay another sub a healthy sum to come in and finish the job. Some subs may not even guarantee work they didn't do completely. *NEVER PAY SUBS FOR MORE THAN THE WORK THAT HAS ALREADY BEEN DONE.* Give them a reason to come back.

SPECIFICATIONS
Don't count on getting anything you don't ask for in writing. *REMEMBER: NOT WRITTEN, NOT SAID.* A list of detailed specifications reduces confusion between you and your sub and will reduce call backs and extra costs. Different subs have different ways of doing things. Make sure they do it YOUR way.

PAPERWORK
Remember that many subs live from week to week and despise paperwork. You will have dif-
ficulty getting many subs to present bids or sign affidavits. Many are stubborn and believe that "their word is their bond". You must use discretion in requiring these items. Generally your more skilled trades such as HVAC, Electrical, and Plumbing will be more business-like and will cooperate more fully with your accounting procedures. These trades are the most important to get affidavits from because they purchase goods from other suppliers that become part of your house. Labor-only subs need not sign affidavits. Make sure that all your subs provide you with an invoice for work done and have them sign it *PAID IN FULL.* This practice will save you many headaches later.

LICENSES
Depending upon who you have work for you, you may want them to furnish a business license. If required, make sure that plumbers, electricians, HVAC subs and other major subs are licensed to work in your county; if not, have them pay to obtain the license; or get another sub.

ARRANGEMENTS FOR MATERIALS
Material arrangements must be agreed upon in a signed, written specification. If subs are to pay for all materials or some of the materials, this must all be spelled out in detail. If a sub is supposed to supply materials, make sure he isn't billing the materials to your account, expecting you to pay for them later. Refer to Lien Waivers in the Legal section. If you are to supply the materials, make sure they are at the site ahead of time and that there is sufficient material to do the job. This will be appreciated by the subs and will generate additional cooperation in the future.

KEEP YOUR OPTIONS OPEN
Don't count on one excavator, framing crew, plumber or anyone for that matter. Have at least one or two back-up subs who can fill in when the primary doesn't show up or does not please you. It is best to be up front with your primaries; let them know that you plan to count on them and expect them to live up to their word. If they can't be there to do the work, skip to your second string subs.

BE FLEXIBLE WITH TIME
Because of variables in weather, subs and their work, expect some variation in work schedule.

Chances are, they are doing four or five other jobs at the same time. Expect some problems in getting the right guy at exactly the right time.

EQUIPMENT

Your work specifications should specify that subs are to provide their own tools including extension cords, ladders, scaffolding, power tools, saw horses, etc... Before hiring a sub, make sure he has all the tools. Don't plan on furnishing them. Defective tools you supply can make you liable for injuries incurred when using them. If you don't have sawhorses available for your framing crew and if they didn't bring any, count on them spending their first ten minutes cutting up your best lumber to make a few.

WORKMANS COMPENSATION

Make sure your sub's workmans compensation policy is up to date before the job is started and get a copy for your records. If the sub doesn't have workmans compensation make it clear to him that you intend to withhold a portion of his payment to cover the expense. This can be a sore spot later if you forget to arrange this in advance.

STANDARD BIDDING PROCESS

Most of the steps below will apply to the effort of reaching final written agreements with the subs you will be using. These steps will be referred to in each of the individual trade sections.

BD1 REVIEW your project scrapbook and locate potential subcontractors.

BD2 FINALIZE all design and material requirements affecting sub contractor.

BD3 PREPARE standard specifications for job.

BD4 PREPARE sub-contractor packages (Sub-contractor Information Sheet, Standard Contract Terms, and Specification Sheet).

BD5 CONTACT sub-contractor and discuss plans. Mention any forms to be completed and specify deadline (should be written on forms).

BD6 RECEIVE and evaluate completed bids. Refer to evaluation criterion presented on following page.

BD7 SELECT the best three bids, based on price and your personal assessment of the sub.

BD8 COMPARE bids against budget.

BD9 NEGOTIATE with prospective sub. Ask them what their best "CASH" price is. When you can't get the price any lower, get more service out of the same cash by asking them to do other items in lieu of a cash discount.

BD10 SELECT sub. Make it clear to them that they have been selected and that they now need to commit a specific period of time for your job. Explain that you are fair with money but expect good, timely work according to a written agreement.

BD11 CONTACT subs and tell them when you expect them at the site. Discuss upcoming schedule and keep in touch with them or you will lose them. If the building schedule changes, let the sub know as soon as possible so that he can make other plans. This courtesy will be returned to you when you need the sub at the last minute.

<u>Notes</u>

7. Making Changes to Your Plans

Changes are second nature. Most most people can't even order lunch without changing their minds. We all see things and perceive how they could be made better. That's where a good imagination and careful thought is important. The idea here is to *MAKE CHANGES ON PAPER THEN FREEZE THE DESIGN*. The cost of ripping out walls, adding windows or "that breakfast counter" can be staggering. In many cases, the cost of making a change after construction makes the change economically impractical. Making changes late in the game is a good way to wreck your budget. So be good to yourself; plan ahead and make all those modifications and additions to your plans, while you're still just using an eraser.

Let's look at three examples. Suppose you have decided to have a bay window in the kitchen as opposed to the original idea of a flat window. For sake of simplicity, let's assume the two windows are of the same dimensions.

CHANGE DURING PLANNING
You think the change over and decide that the $110 material differential is well worth the money. You take out your eraser and update you drawings, take-off and contract specifications. Not bad.

CHANGE JUST PRIOR TO INSTALLATION
The flat window has been delivered to the site but is uninstalled. Now you will have to negotiate with the sub-contractor who already signed the contract to install the flat window. The sub may charge you a little extra, that's just part of the game. But now you hope that the material supplier will exchange a bay window for a flat one (provided it is still in new condition). You also pay the $110 differential and a restocking fee which was not in your original budget; this is an out-of-pocket expense. The change will cost you money, but at least you aren't undoing work.

CHANGE AFTER INSTALLATION
The flat window has been installed. NOW you (or your spouse) decide to change to a bay window. This could be expensive for several reasons:

- You may have to damage the flat window to remove it so it may not be returnable.

- You have to purchase the bay window, say $700, out-of-pocket. You or the bank didn't count on this.

- You will have to pay to have the flat window removed and the bay window installed. Pray that you didn't paint or wallpaper yet.

In addition, you have to prepare the change order form. Note that in this case you are at the mercy of the sub-contractor. He knows you want the bay window and may charge you heavily for it. If his price is too high, you have three alternatives:

- Pay the man and try to forget this ever happened.

- Do it yourself. This can be dangerous if you don't know what you're doing. If the sub damages the bay window during installation, its his problem. If you damage it, that Bay window may cost you another $200-300.

- Live with the flat window. (This is frustrating. So long as you live in that home, that flat window will be a constant reminder of a careless mistake).

ADVICE

Think changes over carefully on paper. Really take the time and picture what that drawing represents in real life. Is it what you want?

If you must make changes, decide to act as early as possible. If you make a change, pay for it out-of-pocket. That is, with funds you had not earmarked for the project.

<u>Notes</u>

8. Saving Money on Construction

Aside from the obvious ways to save money on a home, such as physically building yourself and reduction of square footage, other measures can be taken to minimize the cost of your home without compromising on quality. Let's look at a few ways to save more money before, during and after construction.

SAVING MONEY BEFORE CONSTRUCTION

The most effective means here are:

- Shop for prices aggressively

- Plan your home to minimize waste (modular construction)

- Plan for extras later

- Plan your home without extras

SHOP FOR PRICES AGGRESSIVELY

This means not accepting bids as being the best job for the best price until you are satisfied. Use your take-off to help keep prices down. Subs are out there to earn a living; and they will try to get the best price possible. If you say the price is right, it's a done deal. Don't be so anxious to give in. The more you save over your estimate early in the game, the more "cushion" you'll have later for unexpected expenses.

PLAN YOUR HOME TO MINIMIZE WASTE

This means going below the surface. Here is an example:

Carpet normally comes in twelve foot rolls. In estimating your job, a carpet sub will charge you for any and all scrap necessary to carpet designated areas. When possible, plan rooms to be twelve feet wide or twenty-four feet long. This reduces waste significantly.

USE MODULAR CONSTRUCTION

Blueprints are often designed with esthetic purposes in mind. That's O.K., but remember that many materials, lumber especially, are designed around 16", 24" and 48" units. To a reasonable extent, plan your home to be comprised of 24" or 48" "units", eliminating waste and cutting time. Refer to the framing section, which describes this approach in more detail.

PLAN FOR EXTRAS LATER

Almost anyone would like to have all or some of the following items in their home:

- Built-in microwave
- Intercom system
- Central vacuum system
- Fireplace
- Electronic alarm/security systems
- Skylights
- Finished bonus rooms
- Patios and decks
- Swimming pools and hot tubs
- Electric garage door opener
- Crown and other fancy molding
- Wallcoverings
- Sidewalk or other paved area
- Expensive landscaping
- Den paneling and built-in bookcases
- Wet bars (stub in plumbing)
- Paved driveway/garage door
- Gas grill

But these items, and many others, are not essential to getting your Certificate of Occupancy. Don't strain to pay for items you can easily add later. Minimize that mortgage! Smarter individuals learn to live without certain luxuries and add them later when a little cash is available. The trick is to prepare for extras by doing work that will difficult or impossible to add later. If you want a gas grill, run the gas line NOW and cap it. Gas lines, plumbing, and wiring are fairly inexpensive to put in during construction and will wait for you until you are ready to add those extras.

PLAN YOUR HOME WITHOUT EXTRAS

Still, certain other extras can be avoided altogether. These include such items as basements, bay windows, extra brick veneer, an extra room or fireplace, expensive cabinetry, etc... While it may be nice to have them, they may be the difference between a lesser home and no home at all.

Basements, for example, can easily cost you 5 to 7% of the cost of your home. Do you want a basement? Do you need a basement? Many only have one for possible resale purposes. No basement is better than a leaky, damp and smelly one.

SAVING MONEY DURING CONSTRUCTION

The most effective ways here are:

■ Do work yourself

■ Make pre-fabricated substitutions

■ Downgrade

DO WORK YOURSELF

This can be lots of fun. You'll feel (and be) helpful, while saving even more money. In most cases, every hour you spend productively on site can save you 10 to 15 dollars. Great! But be careful. There are two main reasons why you shouldn't get carried away with doing too much of the work yourself:

■ There are many jobs you shouldn't even consider doing. Remember, where special skills are needed, they serve an important purpose: accomplish professional results and save you from injury, expensive mistakes, frustration and wasted time. Certain jobs may only appear easy. You can ruin your health sanding hardwood floors or drywall. You can ruin a den trying to put up raised paneling and crown molding.

■ Work can distract from your more important job of overseeing the work of others. Your most valuable skill is that of BOSS - scheduler, inspector, coordinator and referee. Saving $50 performing a small chore could cost you even more than that in terms of problems caused by oversight elsewhere on the site. Doing lots of work yourself invariably extends your project time table. Let the pros do the job quick.

JOBS TO CONSIDER DOING YOURSELF:
- Paint, sand and fill
- Wallpaper
- Install door and light fixtures
- Install mailbox
- Light landscaping
- LIGHT trimwood
- Clean-up
- Organizing materials and locating them where needed (stack bricks where masons will use them, etc.)

JOBS BETTER LEFT TO PROS:
- Framing
- Masonry
- Electrical
- HVAC
- Roofing
- Cabinetwork
- Plumbing
- All others not listed

MAKE PREFABRICATED SUBSTITUTIONS WHERE POSSIBLE.

Many items, such as fireplaces, can be purchased as pre-fabricated units. This saves you labor costs. When pre-fabricated items are used, make sure the sub knows of this and takes it into consideration in his bid. One-piece fiberglass shower and tub enclosures, attic staircases and pre-hung doors are other examples.

DOWNGRADE IF NECESSARY.

Certain items can make a major impact on cost reduction without making a visible impact on your home. Consider hollow vs. solid core doors, pre-fabricated cabinets vs. custom cabinets, pre-fabricated doors and windows vs. custom doors and windows to name a few of your options. Another example is trim. Trim that will be stained will show all imperfections, so it should be as high a grade as possible. But if you plan to paint the molding, use "finger-joint" grade. This is molding made of lots of little scrap pieces glued together. It costs less, but will serve the purpose.

SAVING MONEY AFTER CONSTRUCTION

Here, savings are small, but every little bit helps:

■ Tax deduction on material sales tax

■ Clean up yourself

■ Move in yourself as early as possible

TAX DEDUCTIONS

Tax deductions are always good news. The tens of thousands of dollars of materials you purchase will most likely have sales tax added. Keep all of your receipts (you should for other reasons anyway) and total them up for your next tax return. However, remember that if you set up a corporation with a tax I.D., you may avoid paying a lot of the state sales tax in the first place.

CLEANING UP
Cleaning up yourself can save you money, too. Normally, a cleaning crew works on a home for hours before it goes on sale. This is work you can do yourself for sure. Treat yourself (and your family) to dinner out after this one. Scraping windows with razor blades, paint touch up and trash hauling will take days.

MOVING IN
Most likely, you will build in the same city (or nearby) where you will live during the project. Intercity moving is relatively easy. Moving yourself can save hundreds of dollars. You can save yourself a little rent if you move in early and "finish up" while you're at home. You should ask the local building inspector about minimum requirements for occupancy.

Notes

9. Safety and Inspections

INSPECTIONS AND INSPECTORS

During the course of your project, you will probably run into two primary types of inspectors:

- County and local officials

- Financial institutions

In general, inspectors are on your side. They are there to help ensure that critical work meets minimum safety and structural standards. If certain work fails inspection, you will be required to correct it, re-schedule the inspection and probably pay a re-inspection fee.

COUNTY AND LOCAL OFFICIALS

A certain number of inspections are normally required by state or county officials during the construction process. When you apply for your building permit, ask for a schedule of required inspections. Although actual inspection points will vary from place to place, your inspections will include some of the critical inspection points listed below in order of occurrence:

- Before pouring concrete footings,

- Before pouring concrete foundation,

- After framing rough-in,

- After electrical, HVAC, and plumbing rough-in,

- After laying sewer or septic tank lines and before filling in,

- Final completion.

FINANCIAL INSTITUTIONS

If you borrowed money from a financial institution for your project, they have an active interest in seeing that the money is being spent properly, as intended, and at a rate such that the money is used up just as the project is being completed. In addition to requiring approval of state and local inspections, bank inspectors will normally want to examine the site each time a loan draw is requested. As your financial insti-

tution sees certain phases of work completed, they will release a certain amount of loan money based on the percent of completion of the project. In this way they will make sure that you have "earned" the right to use their money as intended. Each lender sets up their own "draw schedule" allowing a certain percentage for each construction task.

PRE-INSPECTION

Nobody enjoys unpleasant surprises. Before you have anyone show up for an inspection, make sure that you are ready to pass the inspection. Check the work yourself. Use checklists and your subs. This will insure that the formal inspection turns out to be just a mere formality. This eliminates rejected inspections, embarrassment, re-inspection fees and a bad builder reputation. Inspectors will see you in a favorable light if you are always fully prepared for their inspections. Your major subs - plumbers, HVAC and electricians normally schedule their own inspections, but you should follow up.

GUIDELINES FOR WORKING WITH INSPECTORS

- MAKE sure the work to be inspected is ready for inspection.

- DETERMINE how much advance notice is required and always give proper advance notice.

- NOTIFY the inspector to cancel or reschedule the inspection if work is not ready.

- BE present for the inspection so that you can determine exactly what is incorrect and what must be done to correct it if necessary. If you don't proceed in this manner, you may have to wait for a form letter in the mail and then follow up.

- STAY on good terms with all of your inspectors. Be friendly and courteous. Don't argue when they ask for something to be fixed. They can give you good advice. Cooperate!

PROJECT SAFETY

Few disasters can affect your project more adversely than the death or severe injury of you, one of your subs, a neighbor's child or a building partner on site. As they say, "Project safety is no accident". Hence, project safety should play a MAJOR part in your overall planning and routine supervision. The best way to make sure that serious accidents don't happen is to make safety a big issue with your subs. Remain aware of safety and emphasize this to your workers and site visitors.

You should have builder's risk insurance covering your building site and accidents that may occur there.

SITE SAFETY CHECKLIST

_____ Site safety rules communicated to all subs.

_____ First aid kit available? Familiar with basic first aid practices.

_____ Phone numbers for police, ambulance and firehouse available.

_____ Temporary electrical service grounded.

_____ All electrical tools grounded.

_____ All electrical cords kept away from water.

_____ Warning and danger signs posted in appropriate areas.

_____ Hard hats and steel tipped shoes worn where needed.

_____ Other protective gear available such as goggles, gloves, etc. Always use protective goggles when flying fragments are possible.

_____ You set a good example as a safety-minded individual.

_____ Adequate slope on edges of all ditches and trenches over five feet deep.

_____ Open holes and ditches fenced properly.

_____ Open holes in sub-floor properly covered or protected.

_____ Scaffolding used and secure where needed.

_____ Workers on roof with proper equipment.

_____ Excess and/or flammable scrap not left lying around.

_____ No nails sticking up out of boards or other materials.

_____ Gas cans have spark arrest screens.

_____ Welding tanks shut off tightly when not in use. Stored upright.

_____ Area where welders working checked for smoldering or slow burning wood. Houses have burned down because of this!

_____ Proper clearance from all power lines.

10. Estimating Costs

The cost estimate or "takeoff" will be one of the most time-consuming and important phases of your construction project. Jumping into your first construction project without a realistic estimate of your costs could spell financial disaster. Spend a lot of time working on this one before continuing with your project.

"How in the world do I do a take-off with no previous experience?" A fair question indeed. The information in this section should help you quite a bit. In addition, *Estimating Home Building Costs*, (W.P. Jackson, Craftsman Book Company, 6058 Corte del Cedro, P.O. Box 6500, Carlsbad, CA 92008-0992), is an invaluable and time-saving paperback. Now is the time to study your construction books in detail to get a feel for the structure of the building and the amount of materials used in each phase.

The first "materials list" you may come in contact with, may be included with your construction plans. These lists are usually offered as an extra at an additional price. As reliable documents, these lists are practically worthless. Many will be inaccurate or based on outdated construction techniques. There is no way of telling if the list is accurate or not. The best use of this type of takeoff is as a checklist while running your own figures.

Another important source of information will come from the subcontractors that you plan to use. These subs are familiar with the type of construction in your area and will be very knowledgeable in calculating the amounts of materials needed. In fact, most of your subs will offer to calculate the amount of materials needed for you. This can be very valuable especially for framing because of the sheer number of different materials used. Don't rely solely on your sub's estimate! Make sure to run your own figures to compare against their estimate. Subs have a tendency to overestimate just to be on the safe side. They hate to run out of material and have to sit around while you run to the supplier for reinforcements. If your estimate comes in close to the sub's but a little under, consider it a good starting point. Just make sure that you do add in at least 5-10% waste factor to each estimate.

Another good source of estimating information is your local materials supplier. In previous years, suppliers used to do quite a bit of estimating for their customers. If you can find a supplier that still does this, he is worth his weight in gold.

When figuring your takeoff, use the materials sheet provided as a checklist to make sure that you don't forget any necessary items. As you figure each item, enter it onto your purchase order forms, taking care to separate materials ordered from different suppliers onto different invoices.

EXCAVATION

Excavation is usually charged by the hour of operating time. Larger loaders rent for higher fees. Determine whether clearing of the lot is necessary and whether or not refuse is to be buried on the site or hauled off. Below are approximate times for loader work. Remember, these are only approximations. Every lot is different and may pose unique problems. Your grader may supply chain saw labor to cut up large trees.

One Half Acre Lot:	Sm. ldr.	Lge. ldr.
Clearing trees and shrubs	2-5 hrs.	1-4 hrs.
Digging trash pit	2-4 hrs.	1-3 hrs.
Digging foundation	2-3 hrs.	1-3 hrs.
Grading Building Site	1-5 hrs.	1-3 hrs.
Total	7-17 hrs.	4-13 hrs.

Larger loaders can be an advantage and can be cheaper on the long run, especially if the lot is heavily wooded or has steep topography. If the lot is flat and sparsely wooded, use the smaller loader for the best price.

CONCRETE

When calculating concrete, it is best to create a formula or conversion factor that will simplify calculations and avoid having to calculate everything in cubic inches and cubic yards. For instance, when pouring a 4" slab, a cubic yard of concrete will cover 81 square feet of area. This is calculated as follows:

1 Cubic Yard = 27 Cubic Feet = 46,656 Cubic Inches
1 Sq. Ft. of Concrete 4" thick = 576 Cubic Inches (12"x12"x4")

46,656 Cu. In. / 576 Cu. In. = 81 Sq. Ft. of Coverage

By doing this equation only once, you now have a simple formula for calculating slabs 4" thick that you can use for Driveways, Basement floors, and slab floors. For instance, to determine the concrete needed for a 1200 Sq. Ft. slab simply divide by 81.

1200 Sq. Ft. / 81 = 14.8 Cu. Yds. of concrete

NOTE: Always make sure to add a 5 to 10% waste factor to all calculations.

Use this same principle of creating simple formulas for footings, calculating blocks, pouring concrete walls, etc. Just remember to calculate your own set of formulas based on the building codes in your area.

FOOTINGS

Footing contractors will be hired to dig the footings and to supervise pouring of the footings. Footing subs generally charge for labor only and will charge by the lineal foot of footing poured. Pier holes will be extra. You must provide the concrete. These subs may charge more if they provide the forms. If you use a full service foundation company, they will charge you for a turn key job for footing, wall, and concrete. This makes estimating easy.

Footings generally must be twice as wide as the wall it supports and the height of the footing will be the same as the thickness of the wall. The footing contractor will know the code requirements for your area. Ask him for the dimensions of the footing and then figure an amount of concrete per linear foot of footing. (See the section on calculating concrete.) Here is a typical calculation for supporting an 8" Block wall:

Footing dimension: 8" high x 16" wide
8" x 16" x 12"(1 ft. of footing) = 1536 Cu. In.

46,656 Cu. In. / 1536 Cu. In. = 30.38 lineal ft. of footing per Cu. Yd.

With this size footing, figure one cubic yard of concrete for every 31 lineal feet of footing. Make sure to include 4 extra feet of footing for every pier hole. Add 10% for waste.

CONCRETE FLOORS OR SLABS

Use the conversion factor of 81 sq. ft. for 4" slabs or basement floors. If your slabs must be more or less than 4" make sure to calculate a new conversion factor.

MONOLITHIC SLABS

Break a one piece slab into two components: the slab and the footing sections; and then the figure the items separately.

BLOCK FOUNDATIONS & CRAWL SPACES

Concrete blocks come in many shapes and sizes; the most common being 8"x8"x16". Blocks that are 12" thick are used for tall block walls with backfill in order to provide extra stability. Blocks 8" deep and 12" deep cover the same wall area: .888 sq. ft.

To calculate the amount of block needed, measure the height of the wall in inches and divide by 8" to find out the height of the wall in numbers of blocks. The height of the wall will always be in even numbers of blocks plus a 4" cap block or 8" half block (for pouring slabs). Cap blocks are solid concrete 4"x8"x16" blocks used to provide a smooth surface to build on. If you are figuring a basement wall, take the perimeter of the foundation and multiply by .75 (3 block for every 4 ft.); then multiply this figure by the number of rows of block. To calculate the row of cap block multiply the perimeter of the foundation by .75. Always figure 5-10% waste when ordering block.

If your foundation is a crawl space, your footing is likely to have one or more step downs, or bulkheads, as they are called. Step downs are areas where the footing is dropped or raised the height of one block. This allows the footing to follow the contour of the land. As a result, sections of the block wall will vary in height; requiring more or less rows of block. Take each step down section separately and figure 3 blocks for every 4 feet of wall. Multiply the total number of block by the total number of rows and then add all sections together for the total number of blocks needed.

POURED CONCRETE WALLS

Your poured wall subcontractor will charge by the lineal foot for setting forms. Usually, the price quoted for pouring will include the cost of concrete; but if not, you must calculate the amount of concrete needed. Determine the thickness of the wall from your poured wall sub and find the square foot conversion factor for the number of square feet coverage per cubic yard. Divide this factor into the total square footage of the wall. Example:

8' Poured Concrete Wall x 12" x 12" = 1152 cubic inch per square foot of wall

46656 cubic. inch / 1152 cubic inch = 40.5 square foot per cubic yard of concrete

BRICK

To figure the amount of brick needed, figure the square footage of the walls to be bricked and multiply by 6.75 (There are approximately 675 brick per 100 square feet of wall). Add 5-10% for waste.

MORTAR

Mortar comes premixed in bags of masonry cement which consist of roughly one part portland cement and one part lime. Each bag requires about 20 shovels of sand when mixing. To calculate the amount of mortar needed for brick, figure one bag of cement for every 125 bricks. For Block, figure one bag of cement for every 28 Blocks. Make sure to use a good grade of washed sand for a good bonding mortar. Most foundations will require at least 10 cubic yards of sand.

FRAMING

Estimating your framing lumber requirements will be the most difficult estimating task and will require studying the layout and construction techniques of your house carefully. Be sure to have a scaled blueprint of your home as well as a scaled ruler for measuring. Consult with a framing contractor before you begin for advice on size and grade of lumber used in your area.

This is the time you will want to pull out your books on construction techniques and study

them to familiarize yourself with the components of your particular house. Local Building code manuals will include span tables for determining the maximum spans you are allowed without support. Your blueprints will also list the sizes of many framing members and may include construction detail drawings which can be especially helpful in determining the size and type of lumber needed for framing.

FLOOR FRAMING

Determine the size and length of floor joists by noting the position of piers or beams and by consulting with your framing contractor. Floor joists are usually spaced 16" on center. Joists spaced 12" on center are used for extra sturdy floors. If Floor Trusses are used, figure 24" on center spacing. Calculate the perimeter of the foundation walls to determine the amount of sill and box sill framing needed.

Calculate the square footage of the floor and divide by 32(square footage in a 4x8' sheet of plywood) to determine the amount of flooring plywood needed. If the APA Sturdifloor design is used, order 3/4" or 5/8" Tongue & Groove Exterior Plywood.

Bridging between floor joists is used to reduce twisting and warping of floor members and to tie the floor together structurally. Bridging is calculated by taking the total number of floor joists and multiplying by three. This is the lineal feet on bridging needed for one course of bridging. Most floors will require at least two courses of bridging, so double this figure.

If you are using a basement or crawl space, you will probably be using a steel or wood beam to support the floor members. If floor trusses are used, this item may not be needed since floor trusses can span much greater distances. Consult with your local truss manufacturer.

WALL FRAMING

When calculating wall framing lumber, add together the lineal feet of all interior and exterior walls. Since there is a plate at the bottom of the wall, and a double plate at the tip; multiply wall length by three to get the lineal feet of wall plate needed. Add 10% for waste. For precut wall studs (make sure to get the proper length if precut studs are used. There are many sizes.), allow one stud for every lineal foot of wall and two studs for every corner. Count all door and window openings as solid wall. This will allow enough for waste and bracing.

Headers are placed over all openings in load bearing walls for structural support and are doubled. Add together the total width of all doors and windows and multiply by two. Check with your framing sub for the proper size header.

The second floor can be calculated the same way as the first floor; taking into consideration the different wall layouts, of course. The second floor of a two story house can be figured the same as a crawl space floor with the exception that the joists will be resting on load bearing walls instead of a beam. This will require drawing a ceiling joist layout.

ROOF FRAMING

If roof trusses are to be used, figure one truss for every two feet of building length, plus one truss. If the roof is a hip roof or has two roof lines that meet at right angles, extra framing for bridging must be added.

The roof truss manufacturer should calculate the actual size and quantity of roof trusses at the site or from blueprints to insure the proper fit. Get him to do this and present a bid and material list during the estimating process.

"Stick building" a roof is much harder to calculate and requires some knowledge of geometry to figure all the lumber needed. A "stick built" roof is one that is built completely from scratch. The simplest way of finding the length of ceiling joists and rafters is to measure them from your scaled blueprints. Always round to the nearest greater even length, since all lumber is sold in even numbered lengths.

When calculating ceiling joists, draw a joist layout with opposing joists always meeting and overlapping above a load bearing wall. Figure one ceiling joist for every 16" plus 10% extra for waste.

Rafters are also spaced 16" on center. The length of the rafters can be determined by measuring from blueprints, making sure to allow for cornice overhang. Where two roof lines meet, a valley or hip rafter is necessary. Multiply the

length of a normal rafter by 1.5 to get the approximate length of this rafter.

Gable studs will be necessary to frame in the gable ends. The length of the stud should be equal to the height of the roof ridge. Figure one stud per foot of gable width plus 10% waste. This will be enough to do two gable ends since scrap pieces can be used in the short areas of the gable.

Decking for the roof usually consists of 1/2" CDX exterior plywood. Multiply the length of the rafter times the length of the roof to determine the square footage of one side of the roof. Double this figure to obtain total square footage area of the roof. Divide this figure by the square footage of a sheet of plywood (32). Add 10% waste. This is the number of sheets of plywood needed. Remember the square footage of the roof for figuring shingles.

ROOFING SHINGLES

Roofing shingles are sold in "squares" or the number of shingles necessary to cover 100 sq. ft. of roof. First, find the square footage of the roof, adding 1 1/2 square foot for every lineal foot of eaves, ridge, hip, and valley. Divide the total square footage by one hundred to find the number of squares. Shingles come packaged in one third square packages, so multiply the number of squares by three to arrive at the total number of packages needed.

Roofing felt is applied under the roof shingles as an underlayment and comes in 500 sq. ft. rolls. Divide the total square footage of the roof by 500 and add 20% for overlap and waste to determine the number of rolls needed.

Flashing is required around any chimney or area where two roof lines of different height meet and comes in 50 ft. rolls. Measure the length of the ridge if installing roof ridge vents.

SIDING AND SHEATHING

Multiply the perimeter of the outside walls by the height to obtain the total square footage of outside walls. If gables are to be covered with siding, multiply the width of the gable by the height (This figure is sufficient for both gable ends.) and add this to the square footage of the outside walls. This figure is the total square footage of area to be sided.

Different types of siding are sold in different manners. Sheathing and plywood siding are sold in 4x8' sheets (32 sq.ft.). Divide this figure into the total square footage to determine the number of sheets needed. Add 10% for waste. Lap siding, on the other hand, is sold by squares or 1,000's. Make sure to ask if siding is sold by actual square footage or by "coverage area" (The amount of area actually covered by the siding when applied.) Add 10% for waste or 15% if lap siding is applied diagonally.

Corner trim boards are generally cut from 1x2 lumber. Figure two pieces for every inside and outside corner; the length being equal to the height of the wall being sided.

CORNICE MATERIAL

To estimate cornice material, you must first determine the type or style of cornice and the trim materials to be used. Consult blueprints for any construction detail drawings of the cornice.

The following materials are generally used in most cornices:

Facia Boards - 1x6" or 1x8"
Drip Mold (between facia & shingles) - 1x4"
Soffit - 3/8" exterior plywood
Bed Mold - 1x2"
Frieze mold - 1x8"

Calculate the total linear feet of cornice, including the gables, to find the total lineal feet of each trim material needed. Calculate the square footage of the soffit by multiplying the linear feet of cornice by the depth of the cornice. Divide this by the square footage of a plywood sheet to determine the number of sheets needed.

INSULATION

Calculating the amount of wall insulation is easy - Batts of insulation are sold by square footage coverage. Simply multiply the perimeter of the exterior walls to be insulated by the wall height to determine the total square footage to be insulated.

To calculate the amount of blown-in insulation needed, first determine the type of insulation to be used. Each type; mineral wool, fiberglass, and cellulose, have their own R-factors per inch. This must be known in order to determine the thickness of fill. Then multiply the depth in feet (or fraction thereof) by the square footage of the ceiling area to arrive at the cubic foot volume. Blown in insulation is sold by the cubic foot. If batt insulation is used in the ceiling, it can be figured in the same manner as the wall insulation.

Most insulation contractors will quote a price including materials and labor when installing insulation. Ask the contractor to itemize the amount of insulation used, for comparison with your figures.

DRYWALL

Gypsum wallboard is sold by the sheet (4x10, 4x12, etc.) but is estimated by square footage. To find the total amount needed for walls, multiply the total linear feet of inside and outside walls by the wall height. Make sure to count each interior wall twice, since both sides of the wall will be covered. Count all openings as solid wall and add 10% for waste. Some subs will charge by the square foot for material and labor - then add extras for tray ceilings and other special work.

For the ceiling, simply take the finished square footage of the house. Add 10% for waste and then add this to the wall amount for the total amount of sheetrock needed. If the house has any vaulted or tray ceilings, extra sheetrock and labor must by figured in.

Joint finishing compound comes pre mixed in 5 gallon cans and joint tape in 250 ft. rolls. For every 1,000 sq. ft. of drywall, figure one roll of joint tape and 30 gallons of joint compound.

TRIMWORK

Base molding comes in many styles; with clamshell and colonial being the most common, and will be run along the bottom of the wall in every room. Therefore, the lineal foot figure of walls used to calculate sheetrock is equal to the lineal feet of baseboard trim needed. Add 10% for waste.

Make sure to measure the perimeter of any room that will require shoe molding or crown molding. Shoe mold (or quarter round) is usually used in any room with vinyl or wood flooring to cover the crack between the floor and wall. Crown molding around the ceiling may be used in any room but is most common in formal areas such as foyers, living and dining areas. Add 10% for waste.

When ordering interior doors, you may want to purchase prehung doors with preassembled jambs. These doors are exceptionally easy to install and already have the jamb and trimwork attached. Be sure to study the blueprints carefully to determine that the doors ordered open in the right direction and don't block light switches. (There are Right hand and Left hand doors.) To determine which door to order, imagine standing in front of the door; walking in. If you must use your right hand to open the door, it is a right hand door. Most blueprints will have the size and type of each door marked in the opening.

FLOORING

Because of the complexity of laying flooring materials: carpet, hardwood floors, vinyl, or ceramic tile; it is essential to get a flooring contractor to estimate the flooring quantities. An approximate figure would be equal to the square footage of the area covered plus 10% for waste. Carpet and vinyl floor covering are sold by the square yard, so divide by 9 to determine square yardage. Keep in mind however that carpet and vinyl is sold in 12' wide rolls. This dimension may affect the amount of waste in the estimate.

MISCELLANEOUS COVERINGS

This includes ceramic tile for bathrooms or floors, parquet, slate for foyers or fireplaces, and fieldstone for steps or fireplaces. Virtually all these materials are sold by the amount of square footage area covered. Subcontractors who install these items also charge by the square foot; sometimes with material included and sometimes without. Grouts and mortars used in installing these items (with the exception of rock) are premixed and indicate the coverage expected on the container.

PAINT

When hiring a painting contractor, the cost of paint should be included in the cost of the estimate unless you specify otherwise. If you should decide to do the painting yourself, you will need to figure the amount of paint needed. The coverage of different paints vary considerably, but most paints state the area covered on the container. Simply calculate the square footage of the area to be covered and add at least 15% for waste and touchup. Also figure the amount of primer needed. If colors are custom mixed, make sure that you have sufficient paint to finish the job. It is sometimes difficult to get an exact match of a custom color if mixed again at a later date.

CABINETS

Most blueprints include cabinet layouts with the plans. Use this layout if you are purchasing your own cabinets. If the cabinets are installed by a subcontractor, he will take care of the exact measurements. Be sure that painting, staining and installation is included in the price.

WALLPAPER

One roll of wallpaper will safely cover 30 square feet of wall area including waste and matching of patterns. The longer the repeat pattern the more waste. Some European wallpapers vary in coverage. Be sure to check coverage area with the wallpaper supplier. When figuring the square footage of a room to wallpaper, count all openings as solid wall. Also be careful to note if you are buying a single or double roll of wallpaper. Most wallpaper is sold in a roll that actually contains 2 true rolls of wallpaper, or 60 sq. ft. of coverage. Always buy enough paper to finish the job. Buying additional rolls at a later time can cause matching problems if some of the rolls are from a different dye lot. Look for the same lot numbers (runs) on all rolls purchased to assure a perfect match. Use only vinyl wallpaper in baths and kitchen areas for water resistance and cleanup.

MILLWORK AND MISCELLANEOUS

The ordering of windows, lights, hardware, and other specialty items will not be covered here as they are fairly straightforward to calculate.

Other estimates such as heating & air, plumbing, electrical and electrical fixtures must be obtained from the contractors themselves since they provide the materials and labor.

SUBCONTRACTORS

The following section outlines the way most subcontractors charge for their services. Some subs may or may not include the cost of materials in their estimates. Make sure you know your particular sub's policy.

FRAMING--Frame house, apply sheathing, set windows and exterior doors. Charges by the square foot of framed structure. (includes any unheated space such as garage) Extras include Bay windows, chimney chase, stairs, dormers, anything else unusual.

SIDING--Apply exterior siding. Charges by the square of applied siding. Extras: diagonal siding, decks, porches and very high walls requiring scaffolding.

CORNICE--Usually the siding subcontractor. Applies soffit and Facia board. Charges by the linear foot of cornice. Extras:Fancy cornice work, Dentil mold. Sometimes sets windows and exterior doors.

TRIM--Install all interior trim and closet fixtures and sets interior doors. Charges a set fee by the opening or by linear feet of trim. Openings include doors and windows. Extras: stairs, rails, crown mold, mantels, book cases, chair rail, wainscoting, and picture molding.

FOOTINGS--Dig footings, pour and level concrete, build bulkheads (for step downs). Charges by the linear foot of footings. Extras: Pier holes

BLOCK--Lay block. Charges by the block. Extras: Stucco block.

BRICKWORK--Lay brick. Charges by the skid (1000 bricks).

STONEWORK--Lay stone. Charges by the square foot or by bid.

CONCRETE FINISHING--Pour concrete, set forms, spread gravel, and finish concrete. Charges by the square foot of area poured. Extras: Monolithic slab - digging footing.

ROOFING--Install shingles and waterproofs around vents. Charges a set fee per square plus pitch of roof. Ex.- $1 per square over pitch on a 6/12 pitch roof = $7/square. Extras: Some flashing, ridge vents, and special cutout for skylights.

WALLPAPER--Hang wallpaper. Builder provides wallpaper. Charges by the roll. Extras: High ceilings, wallpaper on ceilings, grass cloth.

GRADING--Rough grading and clearing. Charges by the hour of bulldozer time. Extras: Chainsaw operator, hauling away of refuse, travel time to and from site (drag time).

POURED FOUNDATION--Dig and pour footings, set forms, and pour walls. Charges by the linear foot of wall. Extras for bulkheads, more than four corners, openings for windows, doors and pipes.

PEST CONTROL --Chemically treat the ground around foundation for termite protection. Charges a flat fee.

PLUMBING--Install all sewer lines, water lines, drains, tubs fixtures, and water appliances. Charges per fixture installed or by bid. (For instance, a toilet, sink, and tub would be three fixtures.) Installs medium grade fixtures. Extras: Any special decorator fixtures.

HVAC--Install furnace, air conditioner, all ductwork, and gas lines. Charges by the tonnage of A/C or on bid price. Extras: Vent fans in bath, roof fans, attic fans, dryer vents, high efficiency furnaces and compressors.

ELECTRICAL--Install all switches and receptacles; hooks up A/C compressor. Will install light fixtures. Charges by the receptacle. Extras: Connecting dishwasher, disposal, flood lights, door bells.

CABINETRY--Build or install pre-fab cabinets and vanities and apply formica tops. Charges by the linear foot for base cabinets, wall cabinets, and vanities. Standard price includes formica counter tops. Extras: Tile or marble tops, curved tops, pull out shelves, lazy susans.

INSULATION--Install all fiberglass batts in walls, ceilings, floors. Charges by square foot for batts, by the cubic foot for blown-in.

SHEETROCK--Hang sheetrock, tape and finish, stipple. Charges by the square feet of sheetrock. Materials are extra. Extras: Smooth ceilings, curved walls, tray and vaulted ceilings and open foyers.

SEPTIC TANK--Install septic tank. Charges fixed fee plus extra for field lines.

LANDSCAPING--Level with tractor, put down seeds, fertilizer, and straw. Charges fixed fee. Extras: Trees, transplanted shrubs, pine straw, bark chips.

GUTTERS--Install gutters and downspouts. Charges by lineal foot plus extra for fittings. Extras: Half round gutters, collectors, special water channeling, and gutters that cannot be installed from the roof.

GARAGE DOOR --Install garage doors. Fixed fee. Extras: Garage door openers.

FIREPLACE--Supply and install prefab fireplace and flue liner. Extras include gas log lighter fresh air vent, and ash dump.

PAINTING AND STAIN--Paint and stain interior and exterior. Charges by square foot of finished house. Extras: High ceilings, stained ceilings, painted ceilings.

CERAMIC TILE--Install all ceramic tile. Charges by the square foot. Extras include fancy bathtub surrounds and tile counter tops.

HARDWOOD FLOOR--Install and finish real hardwood floors. Charges by square foot. Extras include beveled plank, random plank and herringbone

FLOORING--Install all carpet, vinyl, linoleum, and pre finished flooring. Charges by the square yard. Extras include contrast borders and thicker underlayments.

BIDS BEAT ESTIMATES EVERY TIME

Now that you know how detailed an estimate can be, look at the other side of the coin. You can spend a day calculating how many cubic yards of concrete you will need and how many approximate hours of labor it will take; or you can call up a full-service foundation company and nail down a bid. The moral here is to use bids whenever possible in completing your takeoff. The bottom line is what counts; if your estimate comes out to be $600 to tile a floor and your lowest bid is $750, you've obviously missed the boat - and wasted a little time.

COST ESTIMATE SUMMARY

	EST. COST	ACTUAL COST	OVER/UNDER
LICENSES AND FEES			
PREPARATION OF LOT			
FOOTINGS			
FOUNDATION			
ADDITIONAL SLABS			
MATERIALS (PACKAGE)			
FRAMING LABOR			
TRIM LABOR			
ROOFING			
HVAC			
ELECTRIC			
INTERIOR FINISH (SHEETROCK)			
PAINTING			
FLOOR COVERING			
WALKS, DRIVES, STEPS			
LANDSCAPING			
CABINETS AND VANITIES			
APPLIANCES AND LIGHTS			
INSULATION			
CLEANING			
DEVELOPED LOT COST			
CONSTRUC. CLOSING AND INTEREST			
SUPERVISORY FEE			
REAL ESTATE COMMISSION			
POINTS AND CLOSING			
CONTINGENCY (MISC. EXPENSES)			
SEPTIC			
FIREPLACE			
GUTTERS			
TOTAL			

NOTE: Use the Cost Estimate Checklist that follows as a guide when doing your "take-off". Additional blank forms are included in the Appendix for adding items that may not appear on the checklist.

<u>Notes</u>

COST ESTIMATE CHECKLIST

CODE NO. DESCRIPTION	QTY.	MATERIAL UNIT PRICE	TOTAL MAT'L	LABOR UNIT PRICE	TOTAL LABOR	SUBCONTR. UNIT PRICE	TOTAL SUB	TOTAL
LAYOUT								
STAKES								
RIBBON								
ROUGH GRADE								
PERK TEST								
WELL AND PUMP								
Well								
Pump								
BATTER BOARDS								
2 X 4g								
1 X 6								
Cord								
FOOTINGS								
Re-Rods								
2 X 8 Bulkheads								
Concrete								
Footings								
Piers								
FOUNDATION								
Block								
8"								
12"								
4"								
Caps								
Headers								
Solid 8's								
Half 8's								
Mortar Mix								
Sand								
Foundation Vents								
Lintels								
Portland Cement								
WATERPROOFING								
Asphalt Coat								
Portland Cement								
6 Mil Poly								
Drain Pipe								
Gravel								
Stucco								
FOUNDATION								
(Poured)								
Portland Cement								
Wales								
TERMITE TREAT								
SLABS								
Gravel								

TOTALS: ☐   ☐   ☐   ☐

<u>Notes</u>

COST ESTIMATE CHECKLIST (cont.)

CODE NO.	DESCRIPTION	QTY.	MATERIAL		LABOR		SUBCONTR.		TOTAL
			UNIT PRICE	TOTAL MAT'L	UNIT PRICE	TOTAL LABOR	UNIT PRICE	TOTAL SUB	
	6 Mil Poly								
	Re-Wire								
	Re-Rods								
	Corroform								
	Form Boards								
	2 X 8								
	1 X 4								
	Concrete								
	Basement								
	Garage								
	Porch								
	Patio								
	AIR CONDITIONING								
	House								
	FRAMING SUB FL.								
	I-Beam								
	Steel Post								
	Girders								
	Scab								
	Floor Joist								
	Bridging								
	Glue								
	Plywood Sub Fl.								
	5/8" T&G								
	3/4" T&G								
	1/2" Exterior								
	Joist Hangers								
	Lag Bolts								
	Sub-Floor								
	FRAMING-WALLS/PARTITIONS								
	Treated Plate								
	Plate								
	8' Studs								
	10' STuds								
	2 X 10 Headers								
	Interior Beams								
	Bracing								
	1 X 4								
	1/2" Plywood								
	Sheathing								
	4 X 8								
	4 X 9								
	Laminated Beam								
	Flitch Plate								
	Bolts								
	Dead Wood								

TOTALS:

<u>Notes</u>

COST ESTIMATE CHECKLIST (cont.)

CODE NO. DESCRIPTION	QTY.	MATERIAL UNIT PRICE	TOTAL MAT'L	LABOR UNIT PRICE	TOTAL LABOR	SUBCONTR. UNIT PRICE	TOTAL SUB	TOTAL
FRAMING-CEILING/ROOF								
Ceiling Joist								
Rafters9								
Barge Rafters								
Beams								
Ridge Beam								
Wind Beam								
Roof Bracing Matl.								
Ceiling Bracing								
Gable Studs								
Storm Anchors								
Decking								
3/8" CDX Plywood								
1/2" CDX Plywood								
2 x 6 T&G								
Plyclips								
Rigid Insulation								
Felt								
15 #								
40 #								
60 #								
FRAMING-MISC.								
Stair Stringers								
Firing-in								
Chase Material								
Purlin								
Nails								
16DCC								
8DCC								
8" Glv.Roof.								
1 1/2" Glv. Rf.								
Concrete								
Cut								
ROOFING								
Shingles								
Ridge Vent								
End Plugs								
Connectors								
Flashing								
Roof to Wall								
Roll								
Window								
Ventilators								
TRIM-EXTERIOR								
Doors								
Main Entrance								

TOTALS:

Notes

COST ESTIMATE CHECKLIST (cont.)

CODE NO. DESCRIPTION	QTY.	MATERIAL UNIT PRICE	TOTAL MAT'L	LABOR UNIT PRICE	TOTAL LABOR	SUBCONTR. UNIT PRICE	TOTAL SUB	TOTAL
Secondary								
Sliding Glass								
Doors								
Windows								
Fixed Window								
Glass								
Window Frame								
Material								
Corner Trim								
Window Trim								
Door Trim								
Flashing								
Roof to Wall								
Metal Drip Cap								
Roll								
Beams								
Columns								
Rail								
Ornamental Iron								
NAILS								
16d Galv. casing								
8d Galv. casing								
4D Galv. Box								
LOUVERS								
Triangular								
Rectangular								
SIDING								
CORNICE								
Lookout Mat'l								
Facia Mat'l								
Rake mold								
Dentil mold								
Cont. Eave Vent								
Screen								
3/8" Plywood								
DECKS & PORCHES								
MASONRY, BRICK								
Face								
Fire								
MORTAR								
Light								
Dark								
Sand								
Portland cement								
Lintels								
Flue								

TOTALS:

Notes

COST ESTIMATE CHECKLIST (cont.)

CODE NO. DESCRIPTION	QTY.	MATERIAL UNIT PRICE	MATERIAL TOTAL MAT'L	LABOR UNIT PRICE	LABOR TOTAL LABOR	SUBCONTR. UNIT PRICE	SUBCONTR. TOTAL SUB	TOTAL
Flue								
Ash dump								
Clean out								
Log lighter								
Wall ties								
Decorative brick								
Lime putty								
Muriatic acid								
Sealer								
Under-hearth block								
Air vent								
PLUMBING								
Fixtures								
Misc.								
Water line								
Sewer								
HVAC								
Dryer vent								
Hood Vent								
Exhaust vent								
ELECTRICAL								
Lights								
Receptacles								
Switches-single								
Switches-3 way								
Switches-4 way								
Range-one unit								
Range-surface								
Oven								
Dishwasher								
Trash compactor								
Dryer								
Washer								
Freezer								
Furnace-Gas								
Furnace-Elec.								
A/C								
Heat pump								
Attic fan								
Bath fan								
Smoke detector								
Water htr/Elec.								
Flood lights								
Door Bell								
Vent a hood								

TOTALS:

Notes

COST ESTIMATE CHECKLIST (cont.)

CODE NO. DESCRIPTION	QTY.	MATERIAL UNIT PRICE	MATERIAL TOTAL MAT'L	LABOR UNIT PRICE	LABOR TOTAL LABOR	SUBCONTR. UNIT PRICE	SUBCONTR. TOTAL SUB	TOTAL
Switched recept.								
150 amp service								
200 amp service								
Disposal								
Refrigerator								
Permit								
Ground fault								
Fixtures								
FIREPLACE								
Log lighter								
INSULATION								
Walls								
Ceiling								
Floor								
SHEETROCK								
Finish								
Stipple								
STONE								
Interior								
Exterior								
TRIM - INTERIOR								
Doors								
Bifold								
Base								
Casing								
shoe								
Window stool								
Window stop								
Window mull								
1x5 Jamb matl.								
Crown mold								
Bed mold								
Picture mold								
Paneling								
OS Corner mold								
IS Corner mold								
Cap Mold								
False Beam matl.								
STAIRS								
Cap								
Rail								
Baluster								
Post								
Riser								
Tread								
Skirtboard								

TOTALS:

<u>Notes</u>

COST ESTIMATE CHECKLIST (cont.)

CODE NO. DESCRIPTION	QTY.	MATERIAL UNIT PRICE	TOTAL MAT'L	LABOR UNIT PRICE	TOTAL LABOR	SUBCONTR. UNIT PRICE	TOTAL SUB	TOTAL
Bracing								
LOCKS								
Exterior								
Passage								
Closet								
Bedroom								
Bath								
Dummy								
Pocket door								
Bolt single								
Bolt double								
Flush bolts								
Sash locks								
TRIM MISC.								
Door bumpers								
Wedge shingles								
Attic stairs								
Scuttle hole								
1x12 shelving								
Shelf & rod br.								
Rod Socket sets								
FINISH NAILS								
3D								
4D								
6D								
8D								
Closet rods								
Sash handles								
SEPTIC TANK								
Dry well								
BACKFILL								
DRIVES & WALKS								
Form boards								
Rewire								
Rerods								
Expansion joints								
Concrete								
Asphalt								
Gravel								
LANDSCAPE								
PAINT & STAIN								
House								
Garage								
Cabinets								
WALLPAPER								

TOTALS:

<u>Notes</u>

COST ESTIMATE CHECKLIST (cont.)

CODE NO. DESCRIPTION	QTY.	MATERIAL UNIT PRICE	TOTAL MAT'L	LABOR UNIT PRICE	TOTAL LABOR	SUBCONTR. UNIT PRICE	TOTAL SUB	TOTAL
TILE								
Wall								
Floor								
Misc.								
GARAGE DOORS								
GUTTERS								
Gutters								
Downspouts								
Elbows								
Splash blocks								
CABINETS								
Wall								
Base								
Vanities								
Laundry								
GLASS								
SHOWER DOORS								
MIRRORS								
ACCESSORIES								
Tissue holder								
Towel Racks								
Soap Dish								
Toothbrush holder								
Medicine Cab.								
Shower rods								
FLOOR COVERING								
Vinyl								
Carpet								
Hardwood								
Slate or stone								
MISCELLANEOUS								
CLEAN UP								
HANG ACCESSORIES								
FLOOR COVERING								
Vinyl								
Carpet								
Hardwood								
Slate or stone								

TOTALS:

<u>Notes</u>

COST ESTIMATE CHECKLIST

CODE NO. DESCRIPTION	QTY.	MATERIAL		LABOR		SUBCONTR.		TOTAL
		UNIT PRICE	TOTAL MAT'L	UNIT PRICE	TOTAL LABOR	UNIT PRICE	TOTAL SUB	

TOTALS: ☐   ☐   ☐ ☐

Notes

SECTION II:
PROJECT
MANAGEMENT

11. Pre-construction

This section covers each of the trades normally encountered in residential construction. They will be covered in the order in which they normally occur in the construction process. The stages covered are:

- PRE-CONSTRUCTION
- EXCAVATION
- FOUNDATION, WATERPROOFING, and PEST CONTROL
- FRAMING
- ROOFING and GUTTERS
- PLUMBING
- HVAC (Heating, Ventilation and Air Conditioning)
- ELECTRICAL
- MASONRY and STUCCO
- SIDING and CORNICE
- INSULATION and SOUNDPROOFING
- DRYWALL
- TRIM and STAIN
- PAINTING and WALLCOVERING
- CABINETRY and COUNTERTOPS
- FLOORING, TILE and GLAZING
- LANDSCAPING

NOTE: The multitude of combinations and options in a home are as infinite as the ways in which to get them done. The steps outlined on the following pages attempt to cover MOST of the possible steps needed to build MOST homes. For this reason, there are bound to be a number of steps in this manual relating to features you may not have planned. For example, you may not have a basement or require stucco work. Skip steps that do not apply to your project. Likewise, there may be special efforts required in your project that are not covered in this manual. Also, there may be a number of steps you will undertake in special or unconventional circumstances that will not be covered in this book. In many instances, exact order of steps is not critical; use your common sense.

PRE-CONSTRUCTION

Before you run out and start building, there are a number of preparation steps which you should complete. These are somewhat frustrating because they show no visible signs of progress and are often time consuming. There are a lot of forms to fill out and chores to do here. But you will be much better off to get them all out of the way. For most of the steps listed in this section, order of completion is not critical; just get them done.

Give yourself ample time to complete these steps. Rome wasn't built in a day. If you are a first-time builder, it is typical to spend AT LEAST as much time planning your project as you spend executing it. That is, if you plan to have your home completed six months after breaking ground, you should spend AT LEAST six months in planning and preparation prior to breaking ground. Once you break ground, there's no turning back and no time to stop, so you'll be relieved that everything has been thought out ahead of time. This section will cover the items that should be completed prior to breaking ground.

PRECONSTRUCTION: STEPS

PC 1 BEGIN construction project scrapbook. This is where you will begin gathering all sorts of information and ideas you may need later. Start with a large binder or a filing cabinet. Break it into sections such as:

- Kitchen ideas
- Bath ideas
- Floor plans
- Landscaping ideas
- Sub-contractor names and addresses
- Material catalogs and prices

Your "scrapbook" will probably evolve into a small library; a wealth of information used to familiarize yourself with the task ahead and serve as a constant and valuable source of reference for ideas to use in the future.

PC 2 DETERMINE size of home. Calculate the size of home you can afford to build. Refer to the HOUSE PLANS section in this book for details on this step.

PC 3 EVALUATE alternate house plans. Consider traffic patterns, your lifestyle and personal tastes. Refer to the HOUSE PLANS section, presented earlier, for details on evaluating house plans.

PC 4 SELECT house plan or have houseplan drawn. If you use an architect, they will probably charge you a certain fee per square foot. Make sure you really like it before you go out and buy a dozen copies of the blueprints!

PC 5 MAKE design changes. Unless you have an architect custom design your home, there will always be a few things you might want to change - kitchen and kitchen cabinets, master bath, a partition, a door, etc... This is the time to make changes. Refer to the material on making changes in the HOW TO BE A BUILDER and HOUSE PLANS sections.

PC 6 PERFORM material and labor take-off. This will be a time - consuming, but important, step. Refer to the COST ESTIMATING section for details on this step. Get bids to home-in on true costs.

PC 7 CUT costs. After you have determined estimated cost, you may need to pull the cost down. Refer to the SAVING MONEY section for details on reducing costs without reducing quality.

PC 8 UPDATE material and labor take-off. Based on changes you may have made, update and recalculate your estimated costs based on bids.

PC 9 ARRANGE temporary housing. If you own a home and must sell it in order to build, or your current home is too far from your future construction site, consider renting an apartment or home close to your site. If possible, select a place between your site and your office (if you have one). This will greatly facilitate site visits. Since it will take approximately six months to build even the smallest house (unless you are using a pre-fabricated "kit"), you should get at least a six-month lease with a month-to-month option thereafter, or a one-year lease. For first-time, part-time builders, count on ten months to a year before move-in. Delays caused by weather, sub absenteeism, material delays and other problems always seem to add months to the effort. Consider that during the course of the project, you will be paying rent, construction interest and miscellaneous expenses, as well as additional travel costs. Plan extra money to handle this.

PC10 FORM a separate corporation if you want to set your "enterprise" off from your personal affairs. This may give you a better standing among members of the building trade if they feel that they're dealing with a company, instead of an individual. Be advised that your lender may not take kindly to this.

PC11 APPLY for and obtain a business license if required. This may be a legal requirement to conduct business as a professional builder in the county where your site is located. Check with your local authorities for advice.

PC12 ORDER business cards. These will be helpful in opening doors with banks, suppliers and subs. List yourself as president for extra clout.

PC13 OBTAIN free and clear title to land. This must be done before applying for a construction loan. Bankers want to make sure that you have something at stake in this project, too. It is strongly recommended that you only build on land that you own outright.

PC14 APPLY for construction loan. In proper terms, you should be applying for a "construction and permanent loan". This saves you one loan closing since the construction loan turns into a permanent loan when construction is completed.

PC15 ESTABLISH project checking account. This will be used as the single source of project funds to minimize reconciliation problems. Write all checks from this account and keep an accurate balance.

PC16 CLOSE construction loan. This allows you access to funds up to your loan amount. You will pay interest only on money drawn from your reserve. Hence you will not be paying interest on money you have not yet used. Remember, though, that money is paid out as work is completed, based on the judgment of bank officials. Hence, you will need a certain amount of money to begin work (excavation and perhaps some concrete work). Do not begin any physical site work until this is done. Many financial institutions will not even loan you money if you have already started construction prior to contacting them.

PC17 OPEN builder accounts with material suppliers. This will involve filling out forms and waiting for mail replies notifying you of your account number. Read and be familiar with the payment and return policies of each supplier.

PC18 OBTAIN purchase order forms. Look for bound, pre-numbered ones with a carbon copy. You will use these forms for all purchases. Pay only those invoices that reference one of your purchase order numbers. (Make sure to advise suppliers that you are using a purchase order system.)

PC19 OBTAIN building permit. This involves filling out building permit applications with your county planning and zoning department. Build a friendly relationship with ALL county officials. You may need to count on them for things later.

PC20 OBTAIN builder's risk policy. Your construction site is a liability. You, laborers, sub-contractors or neighborhood children could get hurt. Materials may get damaged or stolen. An inexpensive insurance policy is a must. Your financial institution and/or your county officials can point you in the right direction. You should also check with your home and/or auto insurance company for a quote.

PC21 OBTAIN Workman's compensation policy. This is normally charged out as a percentage of the cost of your project. There may be a minimum size policy you can purchase.

PC22 ACQUIRE minimum tools. Certain items are a must. Consider locating some of these items at a pawn shop.

- Good ankle height leather construction boots. Preferably with steel tipped toes. Keep these in your trunk through construction.
- Fiberglass hard hat. Preferably yellow or orange. Scratch your name inside.
- Six foot level. Your level will be a frequently used tool throughout the project. Try to get a used one.
- Strong flashlight.
- Roll of heavy duty cotton twine and blue chalk marker.
- Good all purpose 14 oz. framing hammer.
- Good set of heavy duty gloves.
- 50" or 100" steel tape and 25" tape measure.
- First aid kit.
- UL Approved 50 or 100 foot grounded extension cord and work lamp
- Garden hose
- Large broom or shop vacuum
- Trash can

If you do not own a pickup truck and can't borrow one, try to buy an inexpensive trailer that can be attached to your car. You can sell it after the project is over. You can probably borrow a pickup truck from your subs from time to time.

PC23 VISIT numerous residential construction sites in various stages of completion. Observe the progress, storage of materials, drainage after heavy rainfall and other aspects of those sites.

PC24 DETERMINE need for an on-site storage shack. If you do a lot of the work, this could be helpful.

PC25 NOTIFY all subs/labor. Call up all subs and labor to tell them of the date you plan to break ground. This should be done approximately a week before you plan to break ground. Confirm that they still to plan to do the work and at the agreed-upon price. If you must get a backup sub, this is the time to do it - look on your sub list for "number two".

PC26 OBTAIN compliance bond, if necessary. This is primarily for your protection if you have a builder assist you during construction or build for you.

PC27 DETERMINE minimum requirements for Certificate of Occupancy. As soon as you meet these requirements, you can move into your home while you finish it, avoiding unnecessary rent. During final construction stages, concentrate on meeting these requirements. For example, you can do landscaping and certain other tasks later. Don't move in unless you have finished the rooms you plan to live in immediately.

PC28 CONDUCT spot survey of lot. Locate a good, reliable and inexpensive licensed surveyor. Have them furnish you with a signed survey report. This is a last check to confirm that the planned project and excavation areas are within allowable boundaries. Ask your surveyor to help evaluate water drainage if you are uncertain.

PC29 MARK all lot boundaries and corners with two-foot stakes and bright red tape.

PC30 POST construction signs. You should have at least one "DANGER" or "WARNING" sign. Have one sign for each street exposure if you have a corner lot. You may also want another sign which has your name, "company", and phone number on it. Your lender may have their own sign also.

PC31 NAIL building permit to a tree (that will not be torn down) or a post in the front of the lot visible from the street. You may want to provide a little rain protection for it. You cannot do any work on site until this permit is posted.

PC32 ARRANGE for portable toilet. This is often viewed as a luxury among builders. They rent by the month for a nominal fee that includes maintenance.

PC33 ARRANGE for a site telephone with the phone company. If the site is located far away from anything, this becomes more of a necessity; especially in cases of emergency. If you get one, have the phone company install it with a strong metal case with a lock. You will have to provide a sturdy 4x4 pole or tree to hang it on. Make sure you don't have to pay if it gets stolen. It should be in a place where you can hear the phone ring if you are inside. Specify a loud bell.

12. Excavation

Breaking ground. This is perhaps one of the most exciting moments during your home-building project! This phase actually involves several tasks. You may use your grader four times during the project:

- Clear site and excavate.

- Backfill and smooth out basement floor area.

- Cutting the driveway, rough grade and hauling away trash.

- Finish grade

The excavation sub-contractor is one sub you will pay by the hour. The rates can vary, so shop around. You may be able to get a better rate by hiring excavators to work on "off-hours", such as weekends.

Avoid clearing or excavating when the ground is wet or muddy; it will take more time, be less accurate and will create a mess. Know your lot. Do you have large, heavy trees which will need heavy-duty clearing equipment? Or could you get by with lighter equipment which rents for less?

Remember: The more "horsepower" the higher the hourly rate, but the more it can accomplish in less time. But never use overpowered equipment for the job or you'll end up paying too much. Negotiate hours with the sub. Do they include travel time to and from the site? (Drag time.)

If you want to save money cutting the trees down yourself; go ahead. But, cut them four feet from the ground; the bulldozer needs a good piece of the tree to pull the roots out of the ground. If you don't have a fireplace, consider selling the wood.

EXCAVATION: STEPS

EX1   CONDUCT excavation bidding process on all site preparation work.

EX2   LOCATE underground utilities (gas main, water line, electrical or telephone cables) if they cross your property. You may want to call a locater service or your local utilities for help in locating utility lines. Mark their location with color-coded flags and mark them on a copy of the plat. Don't place flags down more than two or three days in advance. Kids love to pull them up. Your excavator will need to know where the utility lines run. Damaging utilities can be time consuming and expensive.

EX3   DETERMINE where dwelling is to be situated exactly. Your surveyor can help you with this. A landscape architect can help you determine exactly where the home should face from a passive solar and appearance aspect. Mark all corners of the house with stakes so the grader will know where to clear. The grader should dig an extra three feet beyond the corner stakes of the house. For consistency you should have the same setback as the adjacent homes if there are any.

EX4   MARK area to be cleared. Trees to be saved should be marked with red tape or ribbon. Allow clearance for necessary deliveries and several parked vehicles. Now is the time to clear an area for the driveway, patio and septic tank area (if you plan to have one).

EX5   MARK where curb is to be cut (if it is to be cut). The standard curb cut is fourteen feet wide. This allows for a slight flare on each side of the driveway area. In some cases, if you do not cut the curb, you will have a tough bump to deal with getting into your driveway. If it is to be cut later, you can remedy the problem by piling dirt or gravel up at the curb. You may need a variance if you plan to cut more than one curb area - as in the case of a circular driveway on a corner lot.

EX6   TAKE last picture of site area. This is the last chance you have to see it untouched.

EX7   CLEAR site area. Be present for this step. This is normally not something that is worthwhile for you to do because the grader can clear in an hour what would take you a week or more by hand. Cut down necessary trees and brush. Pull stumps out of the ground and remove as necessary. At this point you may wish to mark the foundation area with powdered lime or corner stakes. Do not try to save time by cutting down trees yourself; the grader needs the weight of the tree to pull the stump out of the ground.

EX8   EXCAVATE trash pit. Trash pit location should a minimum of 20 feet from the building or driveway. These can save money normally spent on hauling a large amount of scrap off-site. Some builders don't like them because they attract termites that might eventually be attracted to your home. The area may also settle over time as materials decompose. The pit is also a hazard for young children and adults during construction. In most areas it is unlawful to burn or bury trash.

EX9   INSPECT cleared site. Check for proper building dimensions, scarred trees and thoroughness. Have trash, roots, stumps and other debris been hauled away?

EX10   EXCAVATE foundation/ basement area. You must be present when this step is being performed. To execute this step properly, you should have a transit set up. Take depth measures continuously. This will help you to keep the foundation grade level. You are normally paying this person by the hour unless, of course, you have decided on a firm dollar bid. Make sure he works hard doing exactly what you have in mind. It would be handy to have a copy of the written contract with you. Have the excavator also clear the topsoil off the driveway area. This must be done before the crushed stone is put down on it or the stone will just sink into the loose topsoil.

EX11   INSPECT site and excavated area. Refer to attached inspection sheet and contract specifications.

EX12   APPLY crusher run to driveway area. This will provide an excellent base for a paved drive and allows for easy site access for your subs and supply trucks when the ground is muddy.

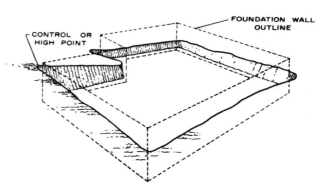

Figure 2. - Establishing depth of excavation.

EX13   SET UP a silt fence. Local ordinances will probably not allow mud run-off from your site to drain into the street or adjacent properties. A silt fence prevents mud from flowing into the street and neighboring yards. Two foot stakes will support the fence. Staple the covering to the stakes with a stapler. Rolls of plastic, mesh, or burlap are available at material suppliers for this purpose, or you may choose to use bales of hay. These bales can be reused later when landscaping.

EX14   PAY excavator. Have them sign a receipt. Pay only for work done up to this point. Give him an idea of when you will be needing him for the backfill work.

BACKFILL

EX15   PLACE supports on interior side of foundation wall to support wall during backfill. Don't backfill until the house is framed in to provide additional support for the foundation walls.

EX16   INSPECT backfill area. Refer to related checklist in this section.

EX17   BACKFILL foundation. Remind excavator not to puncture waterproofing poly with large roots or branches in dirt.

EX18   CUT driveway area.

EX19   COVER septic tank and septic tank line.

EX20 COVER trash pit. Area should be compacted with heavy equipment and covered with at least two feet of dirt.

EX21 PERFORM final grade. Top soil should be spread over top surface. If you have truck loads of top soil hauled in, you may want to get a free soil PH test done by the county extension service. This could save you money in buying fertilizer to compensate for inferior soil.

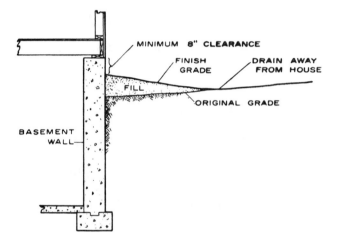

Figure 3. - Finish grade sloped for drainage.

EX22 CONDUCT final inspection. Check all excavation and grading work.

EX23 PAY excavator (final). Have them sign an affidavit.

EXCAVATION: SAMPLE SPECIFICATIONS

» Excavator to perform all necessary excavation and grading as indicated below and on the attached drawing(s) and provide all necessary equipment to complete the job.

» Bid to include all equipment and equipment drag time for two trips.

» Bid to include necessary chain saw work.

» Trees to be saved will be marked with red ribbon. Trees to be saved are to remain unscarred and otherwise undamaged.

» Foundation to be dug to depth indicated at foundation corners. Foundation hole is to be three feet wider than foundation for ease of access. Foundation base not to be dug too deeply in order to pour footings and basement floor on stable base.

» Foundation floor to be dug smooth within two inches of level.

» Excavate trash pit (10' x 12') at specified location as depicted on attached drawing.

» Driveway area to be cleared of topsoil and cut to proper level. All topsoil to be piled where specified.

» All excess dirt to be hauled away by excavator.

» All stumps and trees to be dug up and hauled away. Hardwoods to be cut up to fit fireplace and piled on site where directed.

» Grade level shall be established by the builder who will also furnish a survey of the lot showing the location of the dwelling and all underground utilities.

» Finish grade to slope away from home where possible.

» Backfill foundation walls after waterproofing, gravel and drain tile have been installed and inspected.

» Builder to be notified prior to clearing and excavation. Builder must be on site at start of operation.

EXCAVATION: INSPECTION

CLEARING

_____ Underground utilities left undisturbed.

_____ All area to be cleared is thoroughly cleared and other areas left as they were originally.

_____ All felled trees removed or buried. All remaining trees are standing and have no scars from excavation. Firewood cut and stacked if requested.

EXCAVATION

_____ All area to be excavated is excavated including necessary work space:

» Foundation area (Extra three feet at perimeter)
» Porch and stoop areas
» Fireplace slab
» Crawl space cut and graded with proper slope to insure dry crawl space.

_____ Trash pit dug according to plan if specified.

_____ All excavations done to proper depth with bottoms relatively smooth. Should be within 1" of level.

CHECKLIST PRIOR TO DIGGING AND CLEARING

_____ All underground utilities marked.

_____ All trees and natural areas properly marked off.

_____ Planned clearing area will allow access to site by cement and other large supply trucks. Room to park several vehicles.

_____ All survey stakes in ground.

CHECKLIST PRIOR TO BACKFILL

_____ Any necessary repairs to foundation wall complete.

_____ Form ties broken off and tie holes covered with tar. (Poured wall only.)

_____ All foundation waterproofing completed and correct. This included parging, sprayed tar, poly, etc...

_____ Joint between footing and foundation sealed properly and water tight.

_____ Footing drain tile and gravel installed according to plan.

_____ All necessary run-off drain tile in place and marked with stakes to prevent burial during backfill.

_____ All garbage and scrap wood out of trench and fill area.

_____ Backfill supports in place and secure.

_____ Backfill dirt contains no sharp edges to puncture waterproofing.

CHECKLIST AFTER BACKFILL

_____ All backfill area completely filled in.

_____ Lot smoothed out to a rough grade.

_____ Necessary dirt hauled in and spread.

CHECKLIST AFTER FINAL GRADE COMPLETED

_____ Two to three percent slope or better away from dwelling.

_____ Berms in proper place, of proper height and form.

_____ Water meter elevation correct.

_____ Final grade smooth. Topsoil replaced on top if requested.

_____ No damage to curb, walkways, driveway, base of home, water meter, gutter downspouts, HVAC, splash blocks and trees.

EXCAVATION: GLOSSARY

BACKFILL--Process of placing soil up against foundation after all necessary foundation treatments have been performed. In many instances, foundation wall must have temporary interior bracing or house framed in to support weight of dirt until it has had time to settle.

BATTER BOARD--A pair of horizontal boards nailed to vertical posts set at all the corners of an excavation area used to indicate the desired level of excavation. Also used for fastening taught strings to indicate outlines of the foundation walls.

BERM--Built up mound of dirt for drainage and landscaping.

BUILDERS LEVEL--A surveying tool consisting of an optical siting scope and a measuring stick. It is used to check the level of batter boards and foundation.

BULKHEAD--Vertical drop in footing when changing from one depth to another.

FILL DIRT--Loose dirt. Normally dirt brought in from another location to fill a void. Sturdier than topsoil used under slabs, drives, and sidewalks.

DRAG TIME--Time required to haul heavy excavation equipment to and from the site.

FINISH GRADE--Final process of leveling and smoothing topsoil into final position prior to landscaping.

FOOTING DITCH--Trough area dug to accommodate concrete or footing forms.

FROSTLINE--The depth to which frost penetrates the soil. Footings should always be poured below this line to prevent cracking.

GRADING--Process of shaping the surface of a lot to give it desired contours. See FINISH GRADE.

KERF--Vertical notch or cut made in a batter board where a string is fastened tightly.

PLUMB--The condition when something is exactly vertical to the ground, such as the wall of a house.

PLUMB BOB--A weight attached to a string used to indicate a plumb (vertical) condition.

ROUGH GRADE--First grading effort used to level terrain to approximate shape for drainage and landscaping.

SILT FENCE--A barrier constructed of burlap, plastic, or bales of hay used to prevent the washing away of mud and silt from a cleared lot onto street or adjacent lots.

SETTLING--Movement of unstable dirt over time. Fill dirt normally settles downward as it is compacted by its own weight or a structure above it.

STAKING--To lay out the position of a home, the batter boards, excavation lines and depth(s).

TOP SOIL--Two or three inch layer of rich, loose soil. This must be removed from areas to be cleared or excavated and replaced in other areas later. Not for load bearing areas.

TRANSIT--Similar to a builders level except that the instrument can be adjusted vertically. Used for testing walls for plumb and laying out batter board, and establishing degree of a slope.

<u>Notes</u>

13. Foundation - Waterproofing
Pest Control

The foundation will be one of the most critical stages of construction. Foundation work has one chance to be right - the consequences of poor craftsmanship are permanent. Additionally, this effort will involve individuals from more than one company, which can complicate the job. Hire the best.

This phase of construction requires coordination between a small army of subcontractors and all of them must work with you over a relatively short period of time. The following list briefly describes the sequence of events:

1. Footing subs dig footings.
2. Exterminator treats soil in and around footings to prevent termites.
3. County inspector checks footings before pouring.
4. Footings are poured.
5. Foundation walls are laid or poured. (Basement construction)
6. Plumber and HVAC subs lay water, sewer and gas pipes before slab floor is poured. (Slab construction only)
7. Slab is inspected.
8. Slab is poured by concrete finishers.
9. Basement wall is waterproofed.
10. Basement wall drainage system is installed.

Foundation work can be an exercise in scheduling and diplomacy. Make sure to contact all involved parties well in advance of the pour and plan around rain delays. If possible, cover the foundation area with poly well in advance to prevent rain from ruining your plans. This poly can be used during the pour as the vapor barrier under the slab, if left undamaged.

PEST CONTROL

Below ground, all slab and footing areas must be permanently poisoned prior to laying any gravel, poly or concrete. Poison must be allowed to soak in after application without rain. Rain will reduce or nullify the effectiveness of the pest control treatment. Your pest control treatment can be taken care of in one or two visits depending upon how thorough you want to be. Some builders have a second treatment

applied around the base of the home after all work is done. This is just more insurance; especially if you are building for yourself.

THE CONCRETE SUBS

You will likely deal with several different parties when working with concrete for foundations, slabs and driveways:

- A full service foundation company
- A sub-contractor to dig and pour footings
- Labor to lay concrete block walls
- Concrete finishers to finish concrete slab, patios, walkways and driveways
- Concrete supplier

FULL SERVICE FOUNDATION COMPANY
This subcontractor can be one of the best bargains you will encounter in your project. This sub replaces most of the other foundation subs listed above and may not cost a penny more. A full service foundation company can:

- Lay out the entire foundation for you, including installing batter boards. Only the corners of the foundation need to be marked with stakes. The foundation sub will do the rest.

- Lay out, dig and pour the footings.

- Pour all basement walls.

- Handle all scheduling of inspectors and concrete deliveries.

- Waterproof (tar, gravel, and drain pipe)

Generally, poured foundations are slightly more expensive than other methods; but this cost can be offset by savings in time, labor, hassle and interest paid on your construction loan. However, this sub is only cost effective to use if you plan a full basement.

THE FOOTINGS SUB
The footings sub will normally charge you by the lineal foot of footings required. He will provide either steel or wood forms, but may ask you to supply some of the form material. Form work can cost approximately one third of the total footing job. In many areas of the country

with stable soil, the footings subs can pour directly into the footings trench with no forms required.

CONCRETE BLOCK MASONS

If you do plan to have a basement made of concrete block you'll need this sub. Concrete block masons will likely be a different group than those laying your brick veneer (if you have any). They will charge for labor by the unit (block) laid.

CONCRETE FINISHERS

The concrete finisher may be the footing sub, but not necessarily. He will charge by the square footage finished. Costs will also vary in the type of finishes you want. Slick, rough, smooth, trowel finish or pea gravel are a few of the surfaces that this sub can produce. Don't use a crew that has less than three men - this is tough job. Finishing must be done fast and even faster in hot weather. Concrete can set in half an hour. The finishers will finish the slab and then come back to do drives and walkways later.

CONCRETE SUPPLIER

The concrete supplier will charge by the cubic yard of concrete used. A standard cement truck can hold 8-10 cubic yards. The supplier may charge you a flat rate just to make the trip, so pour and finish as much of the concrete work you can to minimize his trips.

CONCRETE STRUCTURES

The concrete structures that you will use will vary somewhat depending upon whether or not you have a basement. If you plan to have a basement, you will have footings, foundation walls and a foundation floor. If you plan to have no basement, you will have a slab with a footing built into the edge. These structures are described briefly below.

FOOTINGS are what your home rests on. A footing is a solid, sub-grade rectangle of concrete around the perimeter of the dwelling. Its width and thickness will vary depending upon the local building code, soil and weight of materials supported. Footings are normally twice as wide as the wall they are to support. Additional footings are provided for fireplaces and steel column supports.

CONCRETE BLOCKS will be used if you have a basement and you decide not to pour the walls. Concrete blocks are normally hollow units measuring 7 5/8" wide, 7 5/8" high and 15 5/8" long. These are often referred to as 8" x 8" x 16" units because when laid with 3/8" mortar joints, they will occupy a space 16" long and 8" high. Concrete masonry units come in a wide variety of shapes and sizes. Your local building code may have certain guidelines in the use of concrete units.

STEEL REINFORCEMENT is used to strengthen concrete slabs, walls and footings and help it resist the effects of shrinkage and temperature change. Reinforcing comes in two major forms: bars and mesh. Reinforcing bars used for residential construction are normally about 3/8" in diameter. Reinforcing wire, normally with a 6" grid and 1/4" wire is adequate to do the job. For monolithic slabs, mesh is used more than bar steel. Steel reinforcement should be laid before cement is poured, and secured so that no movement will occur during the pouring process. Normally the steel is located 2/3 up from the bottom of a slab and in the center of poured concrete walls. Your concrete sub and concrete supplier will have primary responsibility for proper reinforcement.

FOUNDATION WALLS are made of either concrete blocks or poured concrete.

Concrete Block: This method is normally a little less expensive than a poured wall. However, concrete block walls must be leveled by hand. The concrete block wall must be coated with two 1/4" layers of portland cement below grade level.

Poured Wall: This method, which is growing in popularity, costs a bit more but yields a wall three times as strong as a block wall and more impervious to water. If you are planning to have a brick veneer exterior, you must tell the form sub to provide for a brick ledge in the forms. This is what your brick wall will rest on.

FOUNDATION FLOORS are poured after the footings and walls have been properly cured. Coarse, washed gravel is normally used as an underlayment. The reinforcing is raised to the proper level with rocks or other items. The concrete is normally poured to a depth of four inches. The floor should be finished as close to level as possible with a slight slope toward the drains.

MONOLITHIC SLAB is a combination of footings and foundation floor poured all at once. If fill dirt is used below slab, extra reinforcing bars should be used. The dirt should be well compacted with machinery prior to pouring.

DRIVEWAYS. Driveway slabs are generally 4" thick and should be allowed to cure for 2-3 days before any vehicles are allowed to drive on it. Driveways are not usually installed until trimwork starts.

SIDEWALKS AND PATIOS are simple structures which should only be poured after the surface ground has had ample time to settle. This is usually done when the driveway is poured.

CONCRETE JOINTS

One of the most common reasons for concrete cracking, aside from unstable soil and tree roots is cracking due to expansion and contraction. Expansion and contraction is best controlled by using two types of joints:

(1) Isolation (Expansion Control) Joints. These joints are placed between concrete structures in order to keep one structure's expansion from affecting adjacent structures. Common places used are between:

   » Driveway and garage slab
   » Driveway and roadway
   » Driveway and sidewalk
   » Slabs and foundation wall

(2) Control (Contraction) Joints. These joints are placed within concrete units, such as driveways (normally every 15'), sidewalks (normally every 5' or closer), and patios (normally every 12').

Construction joints, a third type, are temporary joints marking the end of a concrete pour due to delays in concrete supply. If you are lucky, you won't need any of these. Your concrete formsman and/or finisher should be placed responsible for setting all expansion joints and finishers are responsible for all control joints.

WEATHER CONSIDERATIONS

RAINY WEATHER can adversely affect almost any concrete work. Heavy rain can affect poured concrete surfaces and slow the drying process. Schedule your concrete pouring for periods of forecasted dry weather. Have rolls of heavy plastic available if the weather man is wrong.

COLD WEATHER is a challenge. The main problem to prevent here is freezing. Concrete work in temperatures below 40 degrees Fahrenheit becomes difficult. Several methods to protect work include heating water and heating aggregate, covered with tarpaulins over hot pipes. Frozen sub-grade must be heated to ensure that it will not freeze during the curing process. Mix the heated aggregate and water before adding the concrete. Consider using high-early-strength concrete to reduce the time required to protect the concrete from freezing. Once concrete has been poured, and freezing weather is likely, the concrete should be protected either by special heated enclosures or insulation such as bats or blankets.

HOT WEATHER is also a challenge. The main problem to prevent here is rapid drying which induces cracking and reduction in concrete strength. Regardless of the external temperature, concrete should not be allowed to exceed 85 degrees Fahrenheit. This can be achieved by using a combination of the following methods:

■ Cool the aggregate with water before mixing
■ Mix concrete with cold water. Ice can be used in extreme cases
■ Provide sources of shade for materials
■ Work at night or very early in the morning
■ Protect drying concrete from sun and winds which accelerate the drying process. Wet straw or tarpaulins can be used.

Concrete trucks normally hold about 9 cubic yards. Make sure to schedule trucks to provide a constant supply of concrete.

WATERPROOFING

Waterproofing normally applies to foundation wall treatments and drainage tile used to maintain proper foundation water drainage. This is a critical step to ensure that you will have a DRY basement in the many years to come.

WALL TREATMENT
Wall treatments are applied below the finished grade (below ground level). Normally, this treatment consists of hot tar, asphalt or bituminous material applied to the exterior surface. A layer of 6 mil (millimeter) poly (polyvinyl) is then applied to the tar while still tacky. This is an added protection against water seepage. If masonry block walls are used, two 1/4" coats of portland cement are applied to the masonry blocks, extending 6" above the unfinished grade level. The first coating is roughened up before drying, allowing the second coat to adhere properly. Masonry blocks and the first coating of portland cement are dampened before applying coats of portland cement. The second coating should be kept damp for at least 48 hours. The joint between the footing and wall should be filled with portland cement before applying the tar coating. This is called parging and further protects against underground water and moisture.

New waterproofing compounds are available on the market that are superior to asphalt and carry a 10 year warranty if applied correctly. This compound has the ability to bridge cracks in masonry and block and will not degrade with age. It is more expensive than asphalt; however, since wet basements top the list of major homeowner complaints, the extra expense seems well worth it. Ask your waterproofing contractor if he is trained to apply this compound.

DRAIN PIPE
Four-inch perforated drain pipe (plastic preferred over tile) is placed around the perimeter of the foundation with #57 gravel below and 8" to 12" of #57 gravel above. The pipe should be sloped at least 1" every 20 feet. Pipe sections should be spaced 1/4" apart with a strip of tar paper (roofing paper) over each joint. The drain pipe will only serve its purpose if it drains off somewhere. You should make it so that the drain slopes to one corner. You may have a run-off pipe installed to carry the water farther away.

PEST CONTROL: STEPS

PS1   CONDUCT standard bidding process and select pest control sub. Use a company that has been around for at least ten years since you may want to get annual pest control and a warranty. In some cases builders are not charged for the pre treating service. The pest control sub looks for a contract with the homebuyer.

PS2   NOTIFY pest control sub of when to show up for the pre-treat.

PS3   APPLY poison to all footing, slab ground and surrounding area. Since it is hard to test the effectiveness and coverage of treatment, it is advised that someone be present to oversee this critical step. This treatment should only be done if rain is unlikely for 48 hours.

AFTER CONSTRUCTION

PS4   APPLY poison to ground-level perimeter of home. Again, it is advised to have someone oversee this step.

PS5   OBTAIN signed pest control warranty and file it away. Many lenders require a pest warranty at closing to insure proper protection.

PS6   PAY pest control sub and have him sign and affidavit.

CONCRETE: STEPS

CN1   CONDUCT standard bidding process (concrete supplier). Bid will be measured in cubic feet.

CN2   CONDUCT standard bidding process (formwork and finishing) If you use a full service foundation company, they will lay out the foundation, pour footings and walls, and arrange for the concrete. Pricing can get pretty complicated, depending on the height and size of the walls.

CN3   CONDUCT standard bidding process (concrete block masonry).

FOOTINGS

CN4   INSTALL batter boards. Your surveyor might help you with this if you are uneasy or inexperienced. He will also have the proper equipment; such as a builder's level. Here, the area to be excavated is laid out with batter boards and string. Refer to a construction book for instructions on laying out batter boards. If you want help with this, your excavator or someone he knows, can probably give you some help.

CN5   INSPECT batter boards. This is quite critical since your entire project is based on the position of your batter boards and string. This is done with a transit. Solicit the assistance of your surveyor, if possible.

CN6   DIG footings for foundation, A/C compressor, fireplaces and all support columns.

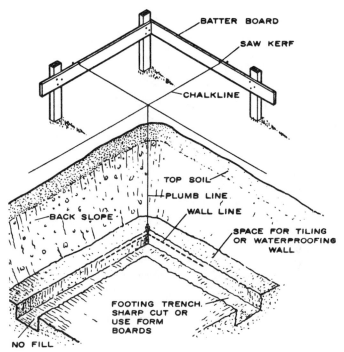

Figure 5. - Establishing corners for excavation and footings.

CN7   SET footing forms as necessary. Be very careful that the footing sub places footings so that walls will sit in the center of the footing. Subs can either use their own forms or use lumber at the site. If the sub asks you to provide forms material, have three times the foundation perimeter in 16 foot 2X4's.

CN8   INSPECT footing forms. Refer to related CHECKLIST in this section. Make sure that the side forms are sturdy as they will have to support lots of wet concrete. Make sure that the tops of the forms are level as they will be used as a guide in finishing the footing surface. Also make sure that the forms are parallel with each other. Footings are normally coated on the inside with some form of oil so that the concrete doesn't stick to them. Check for this. If you plan to have poured walls on your footings, have the forms sub supply the "key" forms which will be pressed into the footings after they are poured.

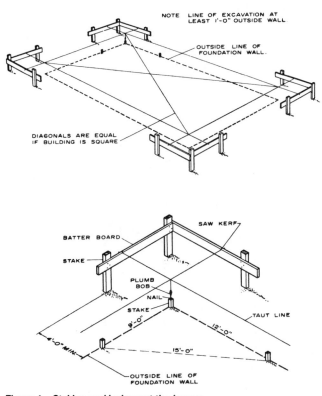

Figure 4. - Staking and laying out the house.

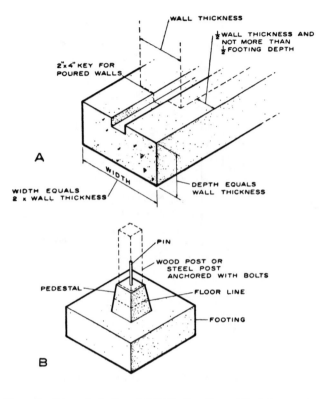

Figure 6. - Concrete footing: A, Wall footing; B, post footing

CN9 SCHEDULE and complete footing inspection. Obtain signed, written approval from local building inspector.

CN10 CALL your concrete supplier to schedule delivery of concrete if you are not using a full service finisher. This should be done at least 24 hours in advance. Make sure good weather is forecast for the next few days. Have rolls of 6 mil poly available in case of rain.

CN11 POUR footings. This should be done as quickly as possible once the footing trenches have been dug so that the soil does not get soft from rain and exposure.

CN12 INSTALL key forms in footings before setting has taken place. Your forms sub should do this. Budget finishers will just cut a groove in the footing with a 2x4.

CN13 FINISH footings as needed. This normally involves simply screeding the top and smoothing out the surface a bit.

CN14 REMOVE footing forms. If subs used your lumber, have them clean and stack it to be used later for bracing. Framers won't use this lumber because it will dull their saws.

CN15 INSPECT footings. Make sure that the footings are level and that there are no visible cracks.

CN16 PAY concrete supplier for footing concrete, unless you are using a full service finisher in which case you will make one payment after the walls have been poured and inspected.

POURED WALL

CN17 SET poured wall foundation forms. These are normally special units made out of steel held together with metal ties.

CN18 CALL concrete supplier in advance to schedule concrete. This step will usually be done by a full service foundation contractor.

CN19 INSPECT poured foundation wall forms. Again, make sure they are level, parallel, sturdy, and oiled.

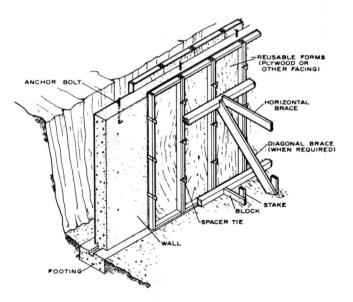

Figure 7. - Forming for poured concrete walls.

CN20 POUR foundation walls. If more than one truck load (10 cu. yds.) is required, make sure to schedule deliveries so that a constant supply of concrete is available for pouring. It is unsatisfactory to have cement setting when wet cement will be poured over it. This can result in a fault crack in the foundation. Make sure that additional loads of concrete are on site or on the way to insure a continuous pour.

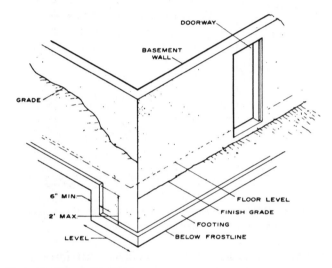

Figure 8. - Stepped footings.

CN21 FINISH poured foundation walls. Again, the tops of the forms serve as screed guides. The surface is smoothed and lag bolts should be set into the surface. Make sure the bolts are at least 4" in the concrete and are standing straight up. Also make sure that the bolts are not in places that will interfere with any studs or door openings. Put wide washers on the sunken end of the lag bolts for greater holding strength. Many builders are omitting lag bolts and just using concrete nails to anchor sill plate to the foundation.

CN22 REMOVE poured foundation wall forms. This should only be done several days after the concrete has had time to set.

CN23 INSPECT poured foundation walls. Check for square, level and visible surface cracks.

CN24 BREAK off tie ends if metal footing forms are used.

CN25 PAY concrete supplier for poured foundation wall concrete.

BLOCK WALL

CN26 ORDER concrete blocks and schedule block masons. Make sure the blocks are placed as near to the work area as possible. Block masons will charge extra if they have to move excessive amounts of block very far.

CN27 LAY concrete blocks. You may want to watch some of this in progress since it will be very hard to correct mistakes when the blocks have set. If you see globs of cement on blocks, wait till it dries to clean it off - it will come off in one piece. Compare against blueprints as the work proceeds.

CN28 FINISH concrete blocks. While mortar is still setting, the masons will tool the joints concave if you like. However, for subgrade work, it is better to have the mortar flush with the blockwork. If any of the block wall is to be exposed, have the block mason stucco the block to provide a smooth, more attractive surface.

CN29 INSPECT concrete block work. Compare work with blueprints. Refer to your specifications and inspection guidelines.

CN30 PAY concrete block masons. Have them sign an affidavit.

CN31 CLEAN up after block masons. Stack or return extra block. Spread left-over sand in drive area. Bags of mortar mix should be protected from moisture.

CONCRETE SLAB

CN32 SET slab forms for basement, front stoop, fireplace pad, garage, etc. Check this against blueprints as they are being erected. If the slab is the floor of the basement, the walls will act as the forms.

CN33 PACK slab sub-soil. Soil should be moistened slightly, then packed with a power tamper. This has got to be a solid surface.

CN34 INSTALL stub plumbing. If the slab is to be the floor of the house, schedule the plumber to set all water and sewer lines under slab area. Ask the plumber to install a foam collar around all sewer lines that protrude out of the slab. This makes final adjustments to the pipe easier during finish plumbing. If the pipe needs to be moved slightly, the foam can be chipped away instead of the concrete. Make sure any resulting gap is resealed with spray foam insulation.

CN35 SCHEDULE HVAC sub to run any gas lines that may run under the slab, such as gas fireplace starters.

CN36 INSPECT plumbing and gas lines against blueprint for proper location. Make sure that plumbing is not dislodged, moved or damaged in subsequent steps.

CN37 POUR and spread crushed stone (crusher run) evenly on slab or foundation floor and garage slab area.

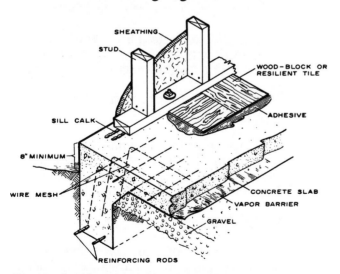

Figure 9. - Combined slab and foundation (thickened edge slab.)

CN38 INSTALL 6 Mil poly vapor barrier. Can be omitted in garage slab.

CN39 LAY reinforcing wire as required. The rewire should lie either in the middle of the slab and garage slab or at a point 2/3 from the bottom. It can be supported by the gravel or rocks.

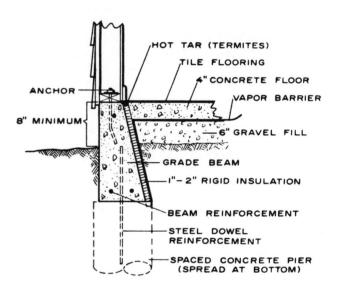

Figure 10. - Reinforced grade beam for concrete slab. Beam spans between concrete piers located below frostline.

CN40 INSTALL rigid insulation around the perimeter of the slab to cut down on heat loss through the slab.

CN41 CALL concrete supplier to schedule delivery of concrete for slab. Confirm concrete type, quantity, time of pour and site location. Finishers should also be contacted.

CN42 INSPECT slab forms. Compare them against the blueprints. Make sure that all necessary plumbing is in place! Check that forms are parallel, perpendicular and plumb as required. Forms should be sturdy and oiled to keep from sticking to concrete. See checklist.

CN43 POUR slab and garage slab. This should be done as quickly as possible. Call it off if rain looks likely. Have 6 mil poly protection available.

CN44 FINISH slab and garage slab. Finishers should put their smoothest finish on all slab work.

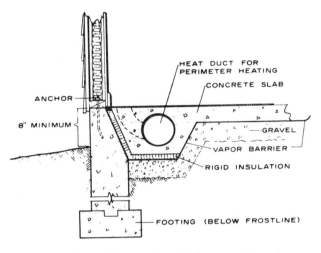

Figure 11. - Full foundation wall for cold climates. Perimeter heat duct insulated to reduce loss.

CN45 INSPECT slab. Check for rough spots, cracks and high or low spots. High and low spots can be check with a hose. See where the water collects and drains. Water should never completely cover a nickel laid flat.

CN46 PAY concrete supplier for slab and garage slab concrete.

DRIVEWAYS AND PATIOS

CN47 COMPACT driveway, walkway and patio sub-soil. This must be a solid surface.

CN48 SET driveway, walkway and patio forms. You should not pour less than a 3" thick driveway. If you did not lay down crushed stone previously, you may want to do it now. If your electrical or gas service crosses the driveway area, it must be installed before this step. Use scrap lumber for forms.

CN49 CALL your concrete supplier to schedule delivery of concrete. Confirm type and quantity of concrete, delivery time and site location. Schedule finishers.

CN50 INSPECT driveway, walkway and patio forms. Pull a line level across forms to determine slope. Forms should be installed so that driveway will slope slightly away from house and toward an area with good drainage.

CN51 POUR driveway, walkway and patio areas. Make sure finishers do not thin concrete for easier working.

CN52 FINISH driveway, walkway and patio areas. Make sure that the finishers put isolation joints between the slab, driveway, patio and walkways. Expansion joints should be put about every 12' in the driveway and every 5' or less in the walkways. Driveway should be formed where it meets the street.

CN53 INSPECT driveway, walkway and patio areas.

CN54 ROPE off drive, walk and patio areas so that people and pets don't make footprints. Put some red tape on the strings so that it can be seen.

CN55 PAY concrete supplier for driveway, walkway and patio concrete.

CN56 PAY concrete finishers and have them sign and affidavit.

CN57 PAY concrete finishers retainage.

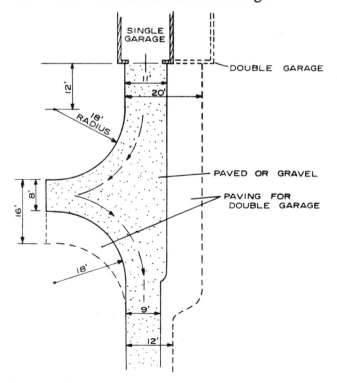

Figure 13. - Driveway turnaround.

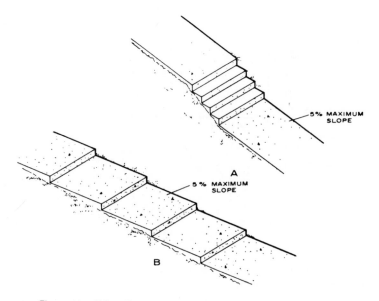

Figure 14. - Sidewalks on slopes: A, Stairs; B, stepped ramp.

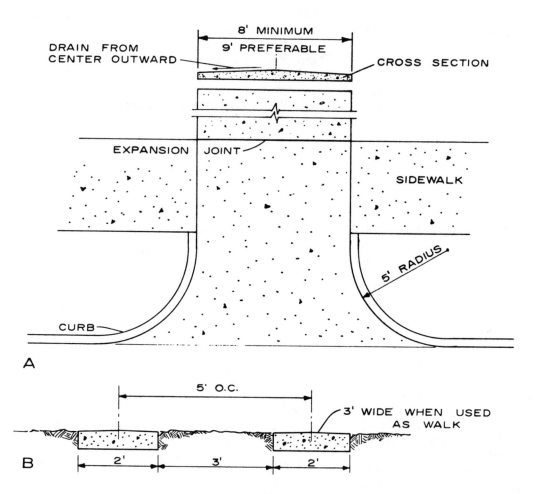

Figure 12. - Driveway details: A, Single-slab driveway; B, ribbon-type driveway.

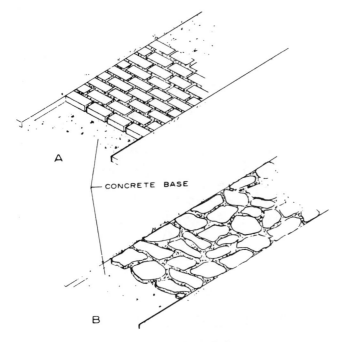

Figure 15. - Masonry paved walks: A, Brick; B, flagstone.

WATERPROOFING: STEPS

WP1 CONDUCT standard bidding process. Invite your full service company to bid.

WP2 SEAL footing and wall intersection with portland cement. Skip if you have no basement.

WP3 APPLY 1/4" portland cement to moistened masonry wall. Roughen surface and allow to dry 24 hours. Perform only if you have a masonry foundation.

WP4 APPLY second 1/4" porland cement to moistened wall. Keep moist for 48 hours and allow to set. Perform only if you have a masonry foundation.

WP5 APPLY waterproofing compound to all below grade walls.

WP6 APPLY black 6 mil poly (4 mil is a bit too thin) to tarred surface while still tacky. Roll it out horizontally and overlap seams.

WP7 INSPECT waterproofing (tar/poly). No holes should be visible.

WP8 INSTALL 1" layer of gravel as bed for drain tile. Optional for crawl spaces, may use even if you have no basement.

WP9 INSTALL 4" perforated plastic drain tile around entire foundation perimeter. Also install PVC run-off drain pipe.

WP10 INSPECT drain tile for proper drainage. Mark the ends with a stake to prevent them from being buried.

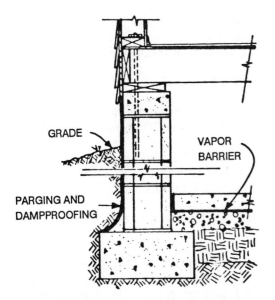

Figure 16. - Dampproofing for basement slab and wall.

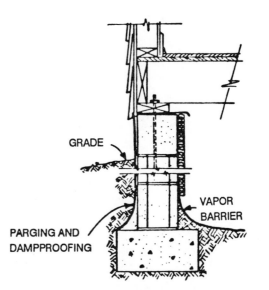

Figure 17. Crawl space dampproofing.

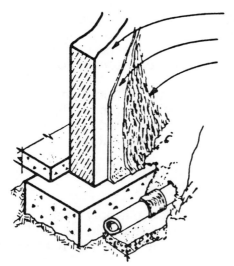

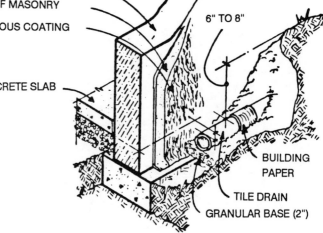

Figure 18. - Drain tiles along side of footing.

Drain tiles on top of footing.

WP11 APPLY 8" to 12" of top gravel (#57) above drain tile. HINT: Surround water spigot area with any excess gravel to prevent muddy mess during construction. Spigot will be used quite often.

WP12 PAY waterproofing sub and have them sign an affidavit.

WP13 PAY waterproofing sub retainage.

CONCRETE: SAMPLE SPECIFICATIONS

GENERAL

- All concrete, form and finish work is to conform to the local building code.

- Concrete will not be poured if precipitation is likely or unless otherwise instructed.

- All form, finishing and concrete work MUST be within one fourth of one inch within level.

- All payments to be made five working days after satisfactory completion of each major structure as seen fit by builder.

CONCRETE SUPPLIER

- Concrete is to be air-entrained ASTM Type I (General Purpose), 3000 psi after 28 days.

- Concrete is to be delivered to site and poured into forms in accordance with generally accepted standards.

- Washed gravel and concrete silica sand to be used.

- Each concrete pour to be done without interruption. No more than one half hour between loads of cement to prevent poor bondage and seams.

PEST CONTROL: SPECIFICATIONS

■ Bid is to provide all materials, chemicals and labor to treat all footing and slab area.

■ Base of home to be sprayed after all construction is complete.

■ Termite treatment warranty to be provided.

CONCRETE: INSPECTION

BATTER BOARDS

____ All batter boards installed:

» Exterior walls
» Piers and support columns
» Garage or carport
» Fireplace slabs
» Porches and entry way
» Other required batter boards

____ All batter boards installed properly:

» Level
» Square
» Proper dimensions according to blueprints
» All strings tight and secure
» All String nails to be marked to prevent movement.

____ All bulkheads well-formed and according to plan. Step-down depth dimension correct.

CHECKLIST PRIOR TO POURING ANY CONCRETE

____ All planned concrete and form work has been inspected by state and/or local officials if necessary prior to pouring.

____ Forms laid out in proper dimensions according to plans. Length, height, depth, etc...

____ Correct relationship to property lines, set-back lines, easements and site plan. (Front, back and all sides).

____ Forms secure with proper bracing.

____ Forms on proper elevation.

____ No presence of ground water or mud.

____ No soft spots in ground area affected.

____ Ground adequately packed.

____ Forms laid out according to specifications: parallel, perpendicular and plumb where indicated.

____ All corners square.

____ Necessary portions straight. No bowing. Wood forms should be nominal 2" thick.

____ Spacers used where necessary to keep parallel forms proper distance apart.

____ Expansion joints in place and level with top of surface. Approximately 1/2" wide.

____ All plumbing underneath or in concrete is installed, complete, inspected and approved.

____ Reinforcing bars bent around corners - no bars just intersecting at corners.

____ Concrete supplier scheduled.

____ Adequate finishers available to do finish work.

____ Any holes under forms filled with crushed stone.

____ Proper concrete ordered.

____ Forms tested for strength prior to pouring.

____ Backfill stable and not in the way. Will not roll onto freshly finished cement or get in finisher's way.

_____ Garage, carport, patio, entrance and porch slabs angled to slop away from home.

_____ Any necessary plumbing has been inspected.

_____ Gravel, re-wire, and 6 mil poly have been placed properly when and where required.

_____ All water and sewer lines installed and protected from abrasions during pouring of concrete.

_____ Finishers have been scheduled.

_____ Crawl space, if any, has been cleared with a rake.

_____ Foam collars placed over all water and sewer pipes where they protrude out of the concrete slab.

GENERAL CHECKLIST AFTER CONCRETE AND BLOCKWORK

_____ Surface checked for level with string level, builder's level or transom.

_____ Corners square using a builder's square.

_____ No cracks, rough spots or other visible irregularities on surface.

_____ Sill bolts vertical and properly spaced. (If used).

FOUNDATION WALL (CONCRETE BLOCK)

_____ Masonry joints properly tooled and even.

_____ Blocks have no major cracks or irregularities.

_____ Cap blocks installed level as top course.

_____ Room allowed for brick veneer above, if any is to be used.

_____ Joint between wall and footing properly parged (slope away from wall) with mortar.

WATERPROOFING: INSPECTION

_____ Portland cement is relatively smooth and even with no sharp edges protruding.

_____ Portland cement completely covers intersection of wall and footings.

_____ Asphalt coating completely covers entire sub-grade area, including intersection of wall and footings.

_____ Black 6 mil poly completely covers entire tar coating and adheres firmly to it. Look for any tears or punctures in poly which need to be patched before backfill. Remove large rocks and roots from backfill area so that poly will not be torn during backfill process later.

_____ Tarred area does not go above grade level or where stucco is to be placed.

_____ Poly secured to stay in place during backfill.

After first heavy rain with roof on: (Pay retainage after this inspection)

_____ No rain in basement.

_____ No wet spots on interior basement walls.

CONCRETE: GLOSSARY

AGGREGATE--Irregular shaped gravel suspended in cement.

CRAWL SPACE--Area below living area where no basement is used.

CRUSHER RUN--Crushed stone, normally up to 2" or 3" in size. Sharp edges. Used for drive and foundation support as a base. Very stable surface, as opposed to gravel, which is not very stable.

CURING--Process of maintaining proper moisture level and temperature (about 73 degrees Fahrenheit) until the design strength is achieved. Curing methods include adding moisture (sprinkling with water, applying wet coverings such as burlap or straw) and retaining moisture (covering with waterproof materials such as polyurethane).

FLOATING--A process used after screeding to provide a smoother surface. The process normally involves embedding larger aggregate below the surface by vibrating, removing imperfections, high and low spots and compacting the surface concrete.

FOOTING--Lowest perimeter portion of a structure resting on firm soil or rock that supports the weight of the structure.

HYDRATION--The chemical process wherein portland cement becomes a bonding agent as water is slowly removed from the mixture. The rate of hydration determines the strength of the bond and hence the strength of the concrete. Hydration stops when all the water has been removed. Once hydrations stops, it cannot be restarted.

LINE--A string pulled tight. Normally to check or establish straightness.

LINTEL--A section of precast concrete placed over doors and windows.

PARGING--Thin coatings (1/4") of mortar applied to the exterior face of concrete block where block wall and footing meet; serving as a waterproofing mechanism.

PLASTIC--Term interchangeable with "wet" as in "plastic cement".

POLY--Polyethylene. A heavy gauge plastic used for vapor barriers and material protection. The accepted thickness is 6 mil.

SCREEDING--The process of running a straight edge over the top of forms to produce a smooth surface on wet (plastic) cement. The screeding proceeds in one direction, normally in a sawing motion.

RE-BAR--Metal rods used to improve the strength of concrete structures.

TERMITE SHIELD--A shield, normally of galvanized sheet metal placed between footing and foundation wall to prevent the passage of termites.

TROWELING--A process used after floating to provide an even smoother surface.

Notes

14. Framing

Framing is one of the most visible signs of progress on the building project. Become close friends with your framing contractor. He will be involved with the project longer than any other contractor. His skills will have a large impact on the quality of your home and your ability to stay on schedule.

Take adequate time and choose your framer carefully. A good framer is worth his weight in 2 X 4's! A good framing crew should consist of a least three people in order to move lumber around efficiently. Check your carpenter's ability to read blueprints. He will be making many interpretations of these plans when framing. Your carpenter should have experience as the master carpenter on several jobs and not just as a helper.

When you feel comfortable with your carpenter's experience proceed by informing him of the quality of work you demand. Ask him to be sure to "crown" all material during construction. What is "crowning"? All lumber has a natural curve to it. Observe the concentric arcs when looking at the end of a board. When framing, the carpenter should be sure to turn all the lumber so that the natural curves of the lumber are in the same direction. When framing horizontal joists, this curve should bow up. The top of the curve is the "crown". Ask your carpenter to lay aside any badly warped or curved lumber to be sent back to the materials supplier or to be used for short framing pieces such as cripples, or dead wood.

Use surfaced, kiln-dried lumber to reduce warping due to shrinkage. Warpage in framing pieces can also be reduced by using a higher grade of lumber and will result in a neater framing job and a happier framing sub. The additional cost is worth it unless you have an unusually careful carpenter willing to spend hours picking out bad lumber. Specify Douglas fir for the best job. Any lumber in contact with the foundation or concrete must be pressure treated pine or "celcure" as it is commonly called. This lumber provides moisture and termite protection between the foundation and the framing of the rest of the house.

PAYMENT OF FRAMING CONTRACTOR

The framing contractor does not usually provide any of his own materials for the job except perhaps nails. It is your responsibility to make sure that all necessary materials are on the site when needed. Carpenters usually charge by the square footage of the area framed plus extras or will appraise the job and supply a total bid price. This includes ALL space covered by a roof, not just heated space. This includes the garage area, enclosed porches, tool sheds, garden houses, etc. The framing charge usually includes setting the exterior doors and windows and installing the sheathing. Special items such as bay windows, stairs, curved stairs, chimney chases, recessed or tray ceilings usually require additional charges.

TYPES OF FRAMING

The two main types of framing used for residential construction are balloon framing and platform framing. For a more complete description of these two methods, refer to a good carpentry manual. Platform framing is the most widely used type and will be the type your carpenter is probably familiar with. In this book, only platform framing will be covered as it is by far the most popular framing method.

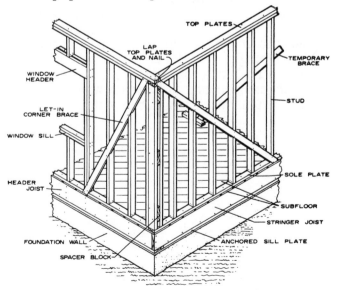

Figure 19. - Wall framing used with platform construction.

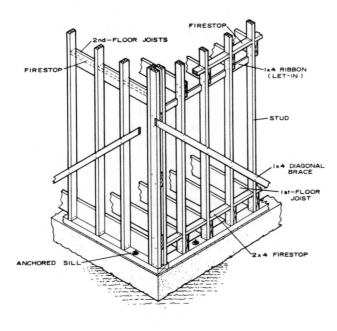

Figure 20. - Wall framing used in balloon construction.

SUB-FLOOR FRAMING

There are two types of floor framing: Joist floors and floor trusses. Joist floors usually consist of 2 x 8's or 2 x 10's spaced 12", 16" or 24" on center with bridging in between. Bridging consists of diagonal bracing between joists to improve stability and to help transfer the load to adjacent joists. Joist floors are by far the most common type in use today, but floor trusses are becoming more popular due to several advantages over the conventional method. A floor truss, like a roof truss, consists of 2 x 4's with diagonal bracing in between which makes the entire unit very strong for its weight.

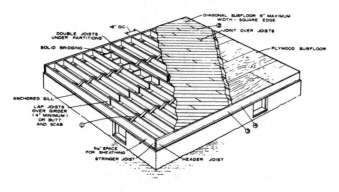

Figure 21. - Floor framing: (1) Nailing bridging to joists; (2) nailing board subfloor to joists; (3) nailing header to joists; (4) toenailing header to sill.

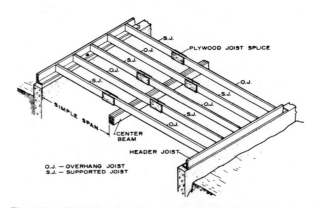

Figure 22. - "In-line" joist system. Alternate extension of joists over the center support with plywood gusset allows the use of smaller joist size.

Floor trusses have the following advantages over standard joists:

■ Because of their added rigidity, floor trusses can span longer distances without the need for beam support.

■ Wiring and heating ducts may be run within the floor truss, eliminating the need to box in heating ducts below ceiling level. This makes the jobs of the plumber, electrician, HVAC, and telephone company much easier. Make sure to point this out to these contractors when bargaining for rates.

■ Floor trusses provide better sound insulation between floors because there is less solid wood for sound to travel through.

■ Floor trusses warp less, squeak less, and are generally stronger than comparable joist floors.

■ Even though floor trusses are a bit more expensive, their advantages can simplify a construction project and can save money in the long run. Your supplier for these units will be the same as for roof trusses.

Figure 23.

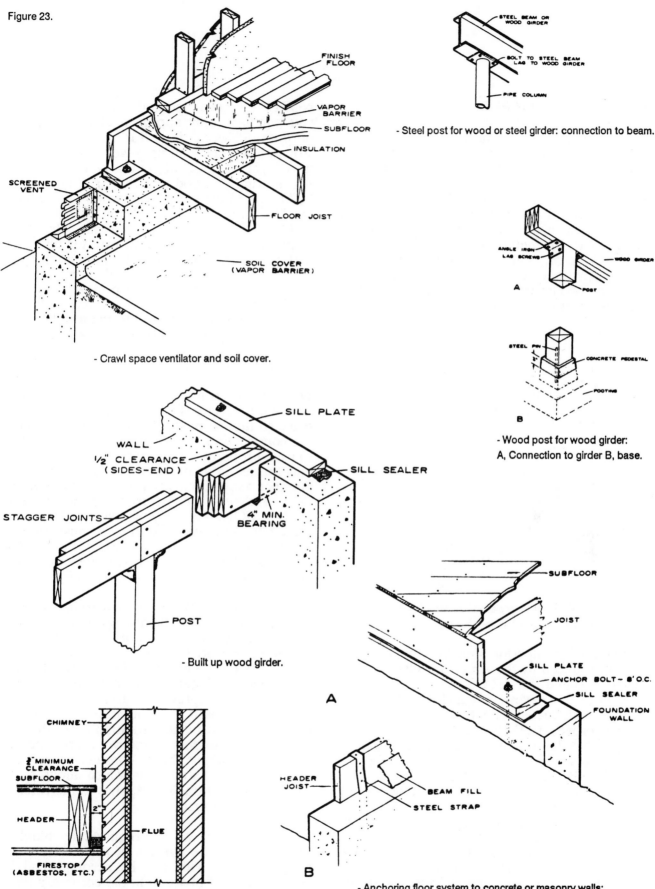

- Crawl space ventilator and soil cover.

- Steel post for wood or steel girder: connection to beam.

- Wood post for wood girder: A, Connection to girder B, base.

- Built up wood girder.

- Clearances for wood construction.

- Anchoring floor system to concrete or masonry walls: A, With sill plate; B, without sill plate.

FLOORING MATERIALS

The flooring design is another area where recent innovations have produced a simpler and superior product. Most standard flooring consists of a sub-floor, usually of 1/2" or 3/4" CDX plywood, and a finish floor which can be 3/4" particleboard or another layer of plywood. The second layer is installed after drying in the structure. If this method is used, be sure to install a layer of asphalt felt (the type used in roofing) or heavy craft construction paper between the two layers. This acts as a sound buffer and reduces floor squeaking.

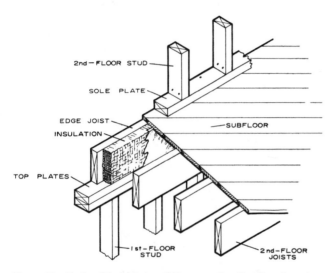

Figure 24. - End-wall framing for platform construction (junction of first -floor ceiling with upper-story floor framing).

A potentially superior flooring method has been introduced by the American Plywood Association called the APA Sturdifloor; a single-layer floor that saves labor and materials. This floor is constructed with one layer of Sturdifloor approved CDX plywood (usually tongue and groove) that is nailed and glued to the floor framing members. Tongue and groove (T & G) plywood has edges that interlock along the two longest edges. This reduces the tendency for boards to flex separately when a load is applied directly to the joint. The act of gluing the plywood to the joists produces an extremely rigid floor and virtually eliminates floor squeaking. This method works extremely well in conjunction with floor trusses and produces a sturdy floor with a minimum of time, hassle, and materials.

NOTE: Whatever method you use, make sure you purchase only approved plywood for this purpose. Plywood used for single layer construction will have an approved APA stamp on it. Ask your supplier about this type of plywood. All flooring plywood must be CDX grade. The "C" refers to the grade of plywood on one side; the "D" refers to the grade on the other side and the "X" refers to exterior grade. This grade of plywood is made with an exterior type glue that is waterproof and will stand up to prolonged exposure to the elements during construction. If you use a two layer floor with particle board as your second layer, do not leave this material exposed to the weather. It is not waterproof and will swell up like a sponge.

WALL FRAMING

Wall framing is a critical stage in construction. You must be sure that everything is in its right place so that surprises do not occur later. During construction, make sure that the walls are spaced properly in relation to plumbing that may be cast in a concrete slab. By the time the plumber may discover a problem you might have a major reconstruction project ahead. DON'T FORGET- If you plan to install one piece fiberglass shower stalls, you must set them in place before putting up the walls! Some fiberglass stalls will not fit through bathroom doors openings.

Make sure to check periodically to be sure that all the carpenter's walls are square. It is easier to get this wrong than you think. Stairs and fireplaces are other sources of potential problems. Check them carefully. Many older carpenters have the habit of installing bridging between studs to prevent warpage. Discourage this practice, as it is an unnecessary waste of lumber and can make insulation more difficult to install and less effective. Also, energy minded individuals should consider using 2 X 6 instead of 2 X 4 stud walls, which allows space for more insulation. This applies to the exterior walls only, of course.

Your carpenter will be required by building code to brace the corners of the structure for added rigidity. This can be accomplished by using a sheet of plywood in the corners, a diagonal wood brace cut into the studs, or a metal strap brace made for that purpose. Although

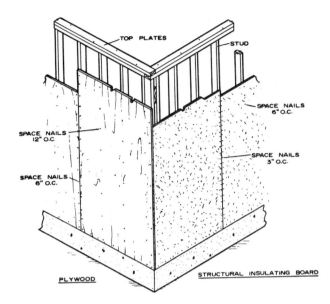

Figure 25. - Vertical application of plywood or structural insulating board sheathing.

more tedious, diagonal bracing or metal strapping is advantageous because it is cheaper and does not interfere with the insulating ability of sheathing. Plywood is not a good insulator.

STUD SPACING

The vast majority of houses in the United States are built with 2 x 4's spaced 16" on center, (apart, or O.C.). But there has been a trend in the past several years to deviate from this norm for cost and energy reasons. Many houses are now constructed with 2 x 6's to allow more room for insulation. This construction change can increase a wall's insulation factor from R-13 to R-19, a significant energy conservation measure. When this method is used, studs are spaced 24" apart. This simplifies construction and reduces the number of studs used by 20-30%. This method is not without it's problems however. The biggest problem is that most doors and windows are designed to be set in walls of 2 x 4 depth. Buying special doors and windows or having custom setting done can cost a considerable sum and can negate many of the cost savings. However, many doors and windows today adjust to accommodate either wall size. Generally, 2 x 6 stud construction is slightly more expensive than conventional construction, but it does provide a significant energy savings. Make sure your carpenter is familiar with the inherent problems of this type of construction before deciding to use it.

Many of the cost savings benefits of 2 x 6 construction can be preserved without any of the headaches by going to 2 x 4 construction spaced 24" on center. This method is approved by all national building codes for single story construction and some local municipalities allow it on two story construction, or at least on the second floor. Check your local building codes to be sure. Spacing studs 24" on center saves time and materials and allows more space for insulation since less studs are used. Ask your carpenter if he is comfortable with this method of construction.

ROOF FRAMING

The two basic types of roof framing consist of stick-built roofs and roof trusses. As with floor trusses, roof trusses have the advantage of being very sturdy for their size and weight. In most house designs, roof trusses require no load-bearing walls between the exterior walls. This gives you more flexibility in designing the interior of your home. Roof trusses are also generally cheaper than stick-built roofs because of significant labor savings in construction. Their only major disadvantage is the lack of attic

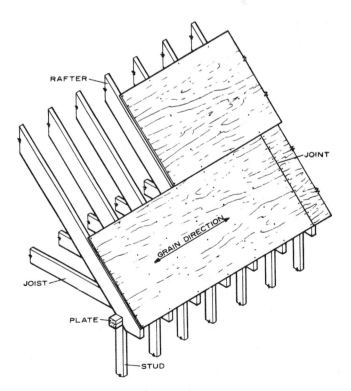

Figure 26. - Application of plywood roof sheathing.

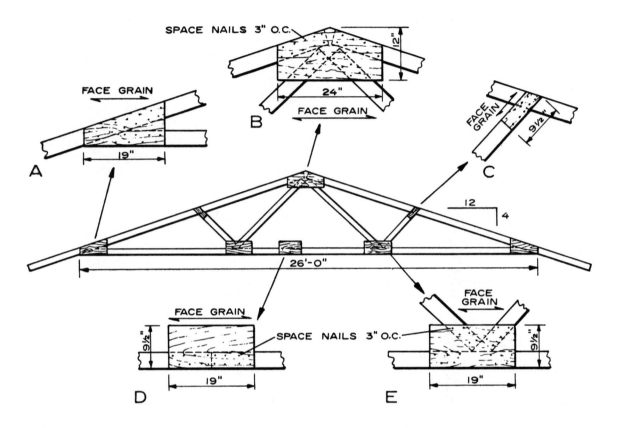

Figure 27. - Construction of a 26-foot W truss: A, Bevel-heel gusset; B, peak gusset; C, upper chord
intermediate gusset; D, splice of lower chord; E, lower chord intermediate gusset.

space as a result of the cross member supports. This can be alleviated somewhat by ordering "space saving" trusses that have the inner cross members angled to provide more attic space. This is called a "W" truss. Stick-built roofs are more appropriate for high pitches, odd shaped or contemporary roof lines.

Your carpenter will install the roof sheathing, which usually consists of 1/2" CDX plywood or pressed cedar panels (cheaper) especially made for this purpose. Request that plywood clips be used in between each sheet of roofing material. These clips are small "H"-shaped aluminum clips designed to hold the ends and sides of the plywood sheets together for added rigidity. This will help prevent warpage and "wavy" roof surfaces.

Due to the popularity of asphalt and fiberglass roofing shingles, this manual relates specifically to roof framing where these two types of roof-

ing are used. Normal roof decking is not used with cedar shake, slate or tile roofs. If you are planning to have such a roof, consult with your framing and roofing subs. When installing the roof sheathing, the carpenter should be aware of the type of roof ventilation to be used. The most convenient and efficient type to use is the continuous eave and ridge vents. The carpenter should leave a 1" gap in the plywood at the roof ridge for installation of the ridge vent, if used.

SHEATHING

Sheathing comes in several different types. Asphalt soaked fiber sheathing is the most common and the cheapest but it has the least insulating value. The next in line would probably be the "blue foam" insulation. It is a good compromise of value and insulation ability. Foil backed urethane is by far the best insulator but also the most expensive. If you plan to stucco, exterior gypsum board is most often used. Talk

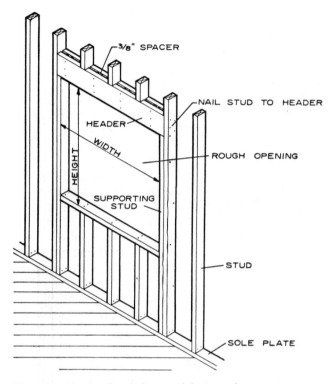

Figure 28. - Headers for windows and door openings.

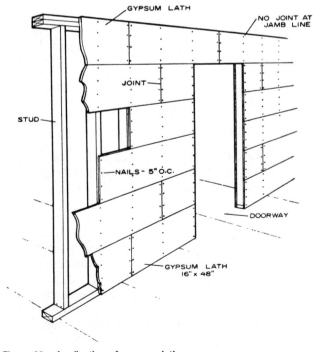

Figure 29. - Application of gymsum lath.

to your local supplier to see what types are available in your area and their insulation ability. When installing the sheathing, your carpenter should take care not to rip or puncture the sheathing. If he does, ask him to replace the defective piece. Pay him only after all defective pieces have been replaced. He should use special nails that have a plate on the shaft that holds the sheathing to the stud without puncturing the sheathing.

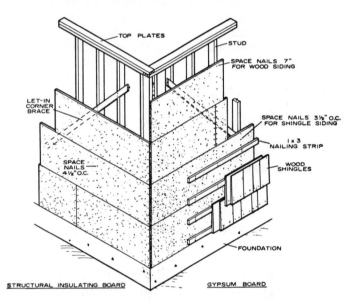

Figure 30. - Horizontal application of 2-by 8-foot structural insulating board or gypsum sheathing.

DOORS AND WINDOWS

After the roof and sheathing have been installed the framing contractor will be ready to install the doors and windows and "dry in" the structure. The proper installation of doors and windows is probably the source of more irritation and hot tempers than anything. A poorly hung door isn't worth much except as an argument starter at the end of a hard day's work. The advent of pre-hung doors has made the job quite a bit easier, but still not foolproof. With your level, check to make sure each door is completely square and that there is an even gap between door and frame from top to bottom. Make sure the door has plenty of room to swell in damp weather (if its wood) and still open freely. Measure under the door to make sure there is plenty of room for subflooring, carpet pad, and carpet without the door rubbing over the carpet when it is installed. The same amount of care should be taken with the windows to guarantee that they will slide freely within their frame. An ounce of prevention is worth a pound of pounding!

OTHER TRADES

Your framing sub must be familiar with the workings of the plumbing and HVAC subs so that he can anticipate for them where it becomes necessary to make room for ductwork, pipes and heating. Make sure his copy of the blueprints indicates the location of furnace and ducting.

FRAMING: STEPS

FR1   CONDUCT standard material bidding process. Find out who has the best package deal on good quality studs, sheathing, press-board, plywood, interior doors (pre-hung or otherwise), exterior doors and windows. Find out their return policy on damaged and surplus material and whether their price includes delivery.

FR2   CONDUCT framing labor bidding process. Ask to see their work and get recommendations. A good crew should have at least three people. Any less and the framing probably won't go as quickly as you would like it. Consider a speed/quality incentive tied to payment.

FR3   DISCUSS and confirm all aspects of framing with the crew. This includes all special angles, openings, clearances and other particulars

FR4   ORDER special materials ahead of time to prevent delays later. This includes custom doors, windows, beams and skylights. This is also a good time to make sure your temporary electric pole works - you will need it for the power saws.

FR5   ORDER and receive first load of framing lumber. Have a nice flat platform ready to lay it on. Have it close to the foundation.

FR6   INSTALL sill felt, sill caulk, or both.

FR7   ATTACH sill plate to the lag bolts imbedded in the foundation, if you use them.

FR8   INSTALL steel or wood support columns in basement.

FR9   SUPERVISE framing process. Check framing dimensions to be sure flooring and wall measurements are in the right place. This is an ongoing process.

FR10   FRAME first floor joists and install subfloor. Place these as close together as you can afford. If you can afford 2 x 10's instead of 2 x 8's, this too will help in giving your floor a more solid feel. Tongue and groove plywood has become a popular sub-floor. Make sure to use exterior grade plywood since it will be exposed to the weather during framing. At bay windows, joists will extend out beyond the wall.

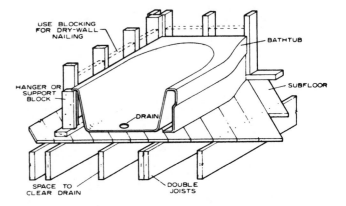

Figure 31. - Framing for bathtub.

FR11   FRAME stairs to basement (if any). In many respects a home is built around staircases. This is a critical step.

FR12   POSITION all large items that must be inserted before framing walls prior to setting doorways and inner partitions. These include bath tubs, modular shower units, HVAC units and oversized appliances. Beware of installing oversized appliances that will not fit through the door when you move or when they have to be replaced.

FR13 FRAME exterior walls/partitions (first floor). Keep track of all window and door opening measurements. Constantly check squareness.

FR14 PLUMB and line first floor. This is critical to a quality job in drywall and trim. This involves setting the top plates for all first floor walls with lapped joints while someone is checking to make sure that all of the angles are plumb and perpendicular. This involves straightening the walls exactly - fine tuning. Walls are braced one by one. Start from one exterior corner and go in a circular direction around the exterior perimeter of the house, adjusting and finishing each wall.

FR15 FRAME second floor joists and subfloor.

FR16 POSITION all large second floor units prior to setting doorways and partitions. Again, these include bath tubs, modular shower units, and other large items.

FR17 FRAME exterior walls/partitions (second floor).

FR18 PLUMB and line second floor. Refer to plumbing/lining first floor.

FR19 INSTALL second floor ceiling joists if roof is stick built.

FR20 FRAME roof. Depending upon the size and complexity of your roof, you may either use pre-fab trusses, or stick build your roof. Chances are, if your home is over 2200 square feet or has a 8/12 or steeper roof, your framers will probably end up stick building it.

FR21 INSTALL roof deck. Plywood sheathing will be applied in 4' x 8' sheets in a horizontal fashion. Each course of plywood should be staggered one half width from the one below it. Your framers should use ply clips to hold the sheets together between rafters.

FR22 PAY first framing payment. About 45% of total cost.

FR23 INSTALL lapped tar paper. This protects your roofing deck and entire structure from rain. At this point, the structure is "dried in" and a major race against the elements is over.

FR24 INSTALL Prefab fireplaces. Install a fireplace with at least a 42" opening and 26" deep. Anything smaller will have problems taking full sized firewood. Prefab fireplaces are usually installed by the supplier.

FR25 FRAME dormers and skylights, if you have them.

FR26 FRAME chimney chases, if you have them.

FR27 FRAME tray ceilings, skylight shafts and bay windows if you have them. Skylight shafts should flare out as it comes into room. This allows more light to enter.

FR28 INSTALL sheathing on all exterior walls.

FR29 INSPECT sheathing. Watch for punctures and gaps in the insulation that can be fixed with duct tape.

FR30 REMOVE temporary bracing supports. Bracing supports installed prior to backfill and those used during the framing process are removed and used as scrap.

FR31 INSTALL exterior windows and doors. The siding and cornice man may do this operation. If you are using pre-hung doors, this becomes an easier process. A pre-hung door comes with its own door frame, but without door knobs and locks. After installing door knobs and locks, the house can be locked up to prevent theft and vandalism. Just make sure to give your subs temporary keys. If your siding and cornice install your windows, this is covered under another step.

FR32 APPLY dead wood. "Dead wood" consists of short pieces of lumber installed in areas that need backing on which to nail drywall.

FR33 INSTALL roof ventilators. This should only be done on the back of the roof.

FR34 FRAME DECKS using pressure treated lumber to prevent rotting and termite infestation.

FR35 INSPECT framing. Hopefully you have been checking on the framing process all along, so there should be no big surprises at this point. This is the LAST chance to get things fixed before you make your last payment to the framing crew.

FR36 SCHEDULE and have your loan officer visit the site to approve rough framing draw.

FR37 CORRECT any problems with framing job. To a large degree, this is an ongoing process that should occur throughout the framing process. The earlier a problem is detected, the easier and cheaper it is to fix. At this point, you have one last, final bargaining tool before making the next payment to the framing crew.

FR38 PAY framing labor. Retain a percentage of pay until all work is completed and final inspection is approved. Remember to deduct any necessary workman's compensation.

FR39 PAY framing labor retainage. Be sure to get a signed affidavit when final payment is made.

FRAMING: SPECIFICATIONS

» All materials to be crowned. (Circular grain goes in the same direction). Where possible, all wall studs have crown pointing the same direction.

» All walls and joists to be framed as specified in the blueprints.

» All joists to be framed 16" OC.

» All vertical wall studs to be framed 16" OC.

» All measurements to be within 1/4" and plumb, perpendicular, and level.

» Cross bridging to be used at all joist midspans to increase strength and stability of floor.

» All lumber in contact with concrete must be pressure treated.

» Double joists to be used under load-bearing partitions and around bathtubs unless truss floors are used.

» Floor joists not to interfere with tub, sink and shower drains.

» 4" overlap on lap joists supported over girders.

» All cuts necessary for HVAC and plumbing work to be braced.

» Hip roof to 10/12 pitch and gable roof to be 12/12 pitch.

» No cuts to be made in laminated beams.

» All chimney chases framed within roof line shall have a cricket installed to allow drainage of excess water.

» Workmans compensation insurance to be provided by builder.

Framing extras include:
» Bay windows
» Stacked bay windows
» Staircases
» Prefab fireplace opening and chase
» Skylights
» Basement stud walls

SHEATHING

» Sheathing to be installed vertically along studs.

» Sheathing nailed every foot along studs.

» All gaps over 1/4" to be taped.

» Skin on sheathing not to be broken by nail heads, hammer or other means.

» Sheathing nails to be used exclusively.

FRAMING: INSPECTION

Framing inspection will be an ongoing process. Errors must be detected as soon as they are made in order for easy correction.

_____ All work done according to master set of blueprints.

_____ No framing deviations exceed standard 1/4" leeway for error.

_____ Vertical walls are plumb.

_____ Opposite walls of rooms parallel.

_____ Horizontal members such as joists, headers and sub-floors are level.

_____ Window and door openings are proper dimensions and are square.

_____ No loose or un-reinforced boards or structures.

_____ Corner bracing in all wall cor-ners.

_____ No room studded without installing large fixtures or appliances that will not fit through door openings later.

_____ All framing is being done to within one inch of on center (o.c.) spacing requested.

_____ Wall corner angles are square. This is especially critical in kitchens (where cabinets and counter tops are expressly designed for 90 degree angles).

_____ Check that vertical studs in stud walls are even and unwarped. This can be done by pulling a string along each stud wall. The string should just touch all studs at the same time.

_____ All joists vertical studs and rafters are crowned. This means that the natural bow of the wood is running in the same direction for all adjacent members.

_____ No excessive scrap.

_____ Adequate lumber is available for framers' use. Have them notify you if shortages are about to occur.

_____ Site measurements match the blueprints. If something is to change, it should be recorded on the blueprints in red.

SUBFLOOR

_____ Specified plywood or other subflooring used. Proper spacing and nailing of sheets.

_____ All sill studs are exterior grade and pressure treated.

_____ All subflooring glued to studs with construction adhesive and nailed adequately.

_____ Bracing between floor joists.

STAIRS

_____ No squeaks or other noise from treads or risers.

_____ Treads deep enough for high-heeled shoes to fit on comfortably.

_____ All risers same height.

_____ All treads same width or as specified.

_____ All framing complete for trimwork to begin.

_____ Attic staircase works properly.

FIREPLACE

_____ Fireplace area headers in proper position for pre-fab or masonry work.

_____ Flue framed with clearances according to building code.

ROOFING: DECK

_____ Plywood sheets staggered from row to row.

_____ Plywood sheets nailed every 8" along rafters or trusses.

_____ Ply clips used on all plywood between supports.

_____ Proper exterior grade of plywood used.

_____ Plywood sheets and roofing felt cut properly for ridge and roof vents.

MISCELLANEOUS: FRAMING

_____ Tray ceiling areas framed as specified.

_____ Tray ceiling angles even.

_____ Skylight openings and skylights according to specifications.

_____ Skylights sealed and flashed.

_____ Access door to crawl space and wall storage areas installed according to specifications.

_____ Pull-down attic staircase installed and fully operable.

SHEATHING

_____ Proper sheathing installed.

_____ Sheathing applied vertically.

_____ Sheathing nails used.

_____ No punctures or other damage to sheathing.

_____ Perimeter of sheets nailed adequately to prevent sheathing from blowing off in wind.

_____ Sheathing joints tight and taped where necessary.

FRAMING: GLOSSARY

BAY WINDOW--Any window that projects out from the walls of the structure.

BEAM--A long piece of lumber or metal used to support a load placed at right angles to the beam - usually floor joists.

LOAD BEARING WALL--Wall that supports the weight of other structural members.

BOARD FOOT--A unit of measurement equal to a 1" thick piece of wood one foot square. Thickness x Length x Width equals board feet.

BRACE--Diagonally framed member used to temporarily hold wall in place during framing.

BRIDGING--Diagonal metal or wood cross braces installed between joists to prevent twisting and to spread the load to adjoining joists.

BUTT JOINT--Junction of the ends of two framing members such as on a sill. Normally a square cut joint.

CASEMENT WINDOW--A window that swings out to the side on hinges.

CORNER BRACES--Diagonal bracing in the corners of a structure used to improve rigidity.

CRICKET--A sloped area at the intersection of a vertical surface and the roof - such as a chimney. Used to channel off water that might otherwise get trapped behind the vertical structure.

CRIPPLE--A short stud used as bracing under windows and other structural framing.

DEAD LOAD--The total weight of walls, floors, and roof bearing on the structure.

DEAD WOOD--Wood used as backing for sheetrock.

DRESSED SIZE--Dimensions of lumber after planing smooth.

DRIED IN--Term describing the framed structure after the roof deck and protective tar paper have been installed.

FACE NAILING--Nailing applied perpendicular to the members. Also known as direct nailing.

GABLE--The vertical section under the sloped part of a roof.

FIRRING--Long strips of wood attached to walls or ceilings to allow attachment of wallboard or ceiling tiles. Firring out refers to adding firring strips to a wall to bring it out further into a room. Furring down refers to using firring strips to lower a ceiling.

GRAIN--The direction of the fibers in the wood. Edge grain wood has been sawed parallel to the growth rings. Flat grain lumber is sawed perpendicular to the growth rings. Studs are flat grain lumber.

GUSSET--A piece of metal or wood used in trusses to connect bracing members together.

HEADER--One or two pieces of lumber installed over doors and windows to support the load above the opening.

JAMB--Exterior frame of a door.

JOIST--A long piece of lumber used to support the load of a floor or ceiling. Joists are always positioned on edge.

KERF--The area of a board removed by the saw when cutting.

LAMINATED BEAM--A very strong beam created from several smaller pieces of wood that have been glued together under heat and pressure.

LINEAL FOOT--A measure of lumber based on the actual length of the piece.

MASTIC--Any pasty material used as a cement or protective coating. Usually comes in caulking tubes or 5 gallon cans.

Notes

15. Roofing and Gutters

Roofing is measured, estimated and bid based on squares (units of 100 square feet as it lays on the roof). Common roofing compositions include asphalt, fiberglass and cedar shakes although special roofs are also comprised of tile, slate and other more expensive materials. Fiberglass is a popular choice due to its favorable combination of appearance, price and durability. Roofing warranties are most often a minimum of 20 years. This pertains to material only.

Asphalt and fiberglass shingles, the most common and inexpensive forms, are of two basic types:

- Strip
- Individual

STRIP SHINGLES are normally 12" tall and 36" wide with two or three tabs to make the units look like several individual shingles when installed. Weights vary from 160 lb. (per square) on the cheap side and up to 280 lb. on the high end. The heavier they are, the more sound and long-lived will be the roof. Lighter shingles also have a tendency to get blown around in heavy wind. Strip shingles normally come in packages of 80 strips. About three packages cover one square of roof. When installed, about 5" of the shingle's height remains exposed, hence a 7" top lap. There are some fiberglass shingles out on the market that resemble slate roofing - from a distance. You just have to look around.

INDIVIDUAL SHINGLES are individual sheets normally measuring 16" X 16" or 12" X 16". Weights vary from 160 lbs. to 330 lbs. Be advised, your roofing sub will probably charge you more to install these than asphalt or fiberglass and they are also not nearly as popular as strip shingles. So make sure you really want them.

ROOFING SUBS should quote you a price with and without material. The steeper your roof pitch, the steeper his price per square installed. Prices start to get steep after about 10/12, which represents the 10 feet of rise for every 12 feet of run (horizontal). Hence, 12/12 is a 45 degree angle. Appeal is growing for steeper and steeper pitches. Make sure the roofing sub has plenty of insurance in addition to your workman's compensation policy. He's one fellow that could really get hurt. If you plan to supply the shingles, make sure they are at the site a day or so early.

NOTE: Since asphalt and fiberglass shingles are by far the most common used, the STEPS on the following pages relate specifically to them. If you use roofing materials other than these two, refer to a handbook and detailed construction methods. Due to the high asphalt content in asphalt and fiberglass shingles, it is highly recommended to apply them is 50 degree Fahrenheit or warmer as they are highly susceptible to cracking in cold weather.

GUTTERS

Gutters are used to channel rain water to downspouts and eventually to planned drainage paths. In some cases, gutters are used more out of appearance than functionality. Typically, gutters are made of galvanized iron or aluminum. The number of downspouts will vary. Your gutter sub-contractor will help you in determining the number and placement of your downspouts. If you can, use underground drain pipes instead of simple splash blocks. Care must be taken that they are not completely buried during the backfill stage. A heavy gauge PVC black tubing should be used.

The primary gutter materials are:

Aluminum--These are good, but you must specify "paint grip" aluminum or the paint will not stick.

Vinyl--This maintenance free alternative is fast becoming a favorite.

Gutter subs normally charge by the lineal feet of gutters and downspouts installed. They may charge you a certain amount for any more than four corners. They will also charge you extra for installing gutter screens and drain pipe.

Get your gutters up as soon as possible to start good drainage and to keep mud from splashing on the side of the house. Gutters shouldn't go on until the cornice is primed and painted and the exterior walls finished (because of downspouts.)

ROOFING: STEPS

RF1   SELECT shingle style, material and color.

RF2   PERFORM standard bidding process for material and labor.

RF3   ORDER shingles and roofing felt (tar paper). This is done only if you have decided to use your roofer for labor only. Check with him to make sure he knows of the brand and how to install them.

RF4   INSTALL 3" metal drip edge on eave nailed 10" on center. Roofing paper will go directly on top of this drip edge (optional.)

RF5   INSTALL roofing felt. This should be done immediately after the roofing deck has been installed and inspected. Your framer will install the tar paper as the last thing before finishing the job. This step protects the entire structure from immediate water damage. Until this step is performed, the project should proceed as quickly as possible.

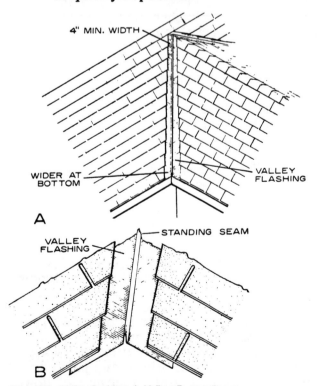

Figure 32. - Valley flashing: A, Valley; B, standing seam.

RF6   INSTALL 3" metal drip edge on rake. This drip edge goes over the roofing paper. Also install aluminum flashing at walls and valleys. (Optional)

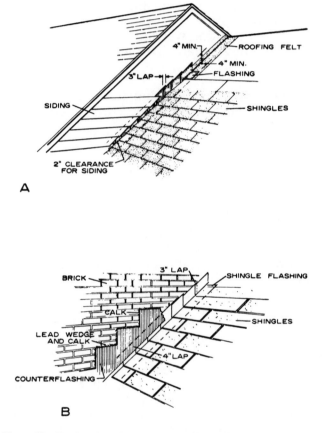

Figure 33. - Roof and wall intersection: A, Wood siding wall; B, brick wall.

RF7   INSTALL roofing shingles. Roofs over 30 feet wide should be shingled starting in the middle. Shorter roofs can be shingled starting at either rake. Starter strips, normally 9" wide, are either continuous rolls of shingle material that start at the eave or doubled shingle strips. Shingles are laid from the eave, overlapping courses up to the highest ridge. Don't install shingles on very cold days; if you have a choice.

RF8   INSPECT roofing. Refer to contract specifications and inspection checklist for guidelines. You will have to get up on the roof to check most of this.

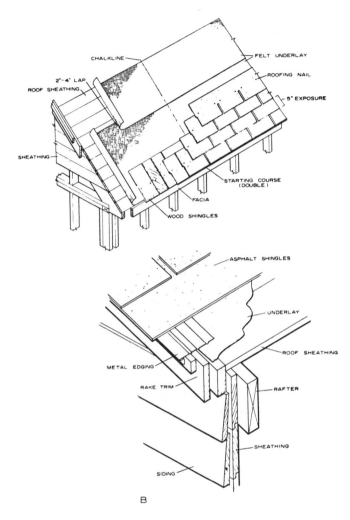

Figure 34. Application of asphalt shingles: A, Normal method with strip shingles; B, metal edging at gable end.

RF9 PAY roofing sub-contractor. Get them to sign an affidavit. Deduct for workmans compensation if applicable.

RF10 PAY roofing sub retainage

GUTTERS: STEPS

GU1 SELECT gutter type and color.

GU2 PERFORM standard bidding process.

GU3 INSTALL underground drain pipe. Flexible plastic pipe is the type used most often. Make sure they are covered properly.

GU4 INSTALL gutters and downspouts.

GU5 INSTALL splashblocks, and/or other water channeling devices.

GU6 INSPECT gutters. Refer to inspection guidelines, related checklist, contract specifications and your building code.

GU7 PAY gutter sub-contractor. Have him sign an affidavit.

GU8 PAY gutter sub retainage.

ROOFING: SAMPLE SPECIFICATIONS

■ Bid is to perform complete roofing job per attached drawings and modifications.

■ Bid to include all materials and labor.

■ Apply 240 lb. 30-year fiberglass shingles (Make/Style/Color) to roof using galvanized roofing nails.

■ Install two thermostatically controlled roof vents on rear of roof as indicated on attached drawings.

■ Install continuous ridge vent per drawing specifications.

■ Roofing felt to be applied and stapled to deck as indicated below:- no. 15 asphalt saturated felt over entire plywood deck.

■ All roofing felt to have a 6" vertical overlap

■ Install 3" galvanized eave and rake drip edges nailed 10" o.c.

■ All work and materials are to conform to or exceed requirements of the local building code and be satisfactory to contractor.

■ Roofer to notify builder immediately of any condition which is or may be a violation of the building code.

■ Nails are to be threaded and corrosion resistant. Either aluminum or hot-dipped galvanized roofing nails to be used depending upon shingle manufacturer recommendations.

■ Nails to be driven flush with shingles.

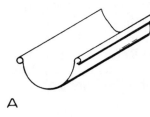
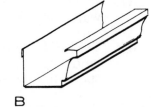
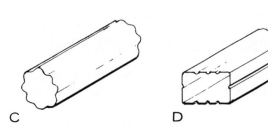

Figure 35. - Gutters and downspouts: A, Half-round gutter; B, formed gutter; C, round downspout; D, rectangular downspout.

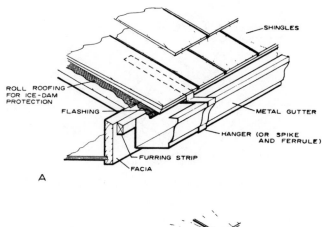

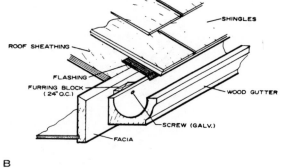

Figure 36. Gutter installation: A, Formed metal gutter; B, wood gutter.

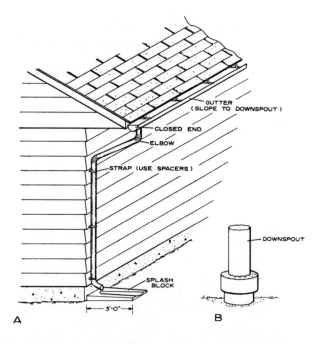

Figure 37. - Downspout installation: A, Downspout with splash block; B, drain to storm sewer.

- Soil stacks to cover all vents. All vents to be sealed to shingle surface with plastic asphalt cement.

- Flashing to be installed at chimney and all roof valleys.

GUTTERS: SAMPLE SPECIFICATIONS

- Bid is to provide and install seamless, paint-grip aluminum gutters around entire lower perimeter of dwelling as specified in attached drawings.

- 5" gutter troughs and 3" corrugated downspouts to be used exclusively.,

- Gutter troughs are to be supported by galvanized steel spikes and ferrules spaced not more than 5 feet apart.

- Downspouts to be fastened to wall every 6 vertical feet.

- Troughs are to be sloped toward downspouts 1 (one) inch for every 12 lineal feet.

ROOFING: INSPECTION

_____ Shingle lines inspected for straightness with string drawn taut. Tips of shingles line up with string.

_____ Even shingle pattern and uniform shingle color from both close-up and afar (at least 80 feet away).

_____ Shingles extend over edge of roofing deck by at least 3".

_____ All roofing nails are galvanized and nailed flush with all shingles. (random inspection).

_____ No shingle cracks visible through random inspection.

_____ Hips and valleys are smooth and uniform.

_____ Roofing conforms to contract specifications and local building codes in all respects.

_____ Shingles fit tightly around all stack vents. Areas are well sealed with an asphalt roofing compound that blends with the shingles.

_____ Drip edges have been installed on eave and rakes.

_____ No nail heads are visible while standing up on the roof.

_____ All shingles lay flat (no buckling).

_____ No visible lumps in roofing due to poor decking or truss work.

_____ Edges of roof trimmed smooth and evenly.

_____ All vents and roof flashing painted proper color with exterior grade paint.

_____ All garbage on roof removed.

_____ Shingle tabs have been glued down if this was specified. (Used only in very windy areas or where very light weight shingles are used).

GUTTERS: INSPECTION

_____ Proper materials used as specified. Aluminum, vinyl, etc...

_____ Proper trough and downspout sizes used.

_____ Downspouts secured to exterior walls.

_____ Adequate number of gutter nails used to support gutters securely.

_____ Water drains well to and through troughs when water from a garden hose or rain water is applied.

_____ No leaks in miters and elbows while water is running.

_____ Water does not collect anywhere in troughs and is completely drained within one minute.

ROOFING AND GUTTERS: GLOSSARY

ASPHALT--Base ingredient of asphalt shingles and roofing paper (felt composition saturated in asphalt base).

BULL--A covering for a wide soil stack so that rain will not enter.

COURSE--A horizontal row of shingles.

COVERAGE--The maximum number of shingles that overlap in any one spot. This determines the degree of weather performance a roof has.

DOWNSPOUT--Vertical member used to bring water from trough to ground level.

DRIP EDGE--Metal flashing normally 3" wide which goes on the eave and rakes to provide a precise point for water to drip from so that the cornice does not rot.

EAVE--Lowest point on a roof that overhangs edge of house.

ELBOW--Trough corner extending outward from roof.

EXPOSURE--The vertical length of exposed shingle (portion not lapped by the shingle above.

FELT--Typical shingle underlayment. Also known as roofing felt or tar paper. See ASPHALT.

FERRULE--Aluminum sleeve used in attaching trough to gutter spike.

FLASHING--Galvanized sheet metal used as a lining around joints between shingles and chimneys, exhaust and ventilation vents and other protrusions in the roof deck. Flashing helps prevent water from seeping under the shingles.

HEAD LAP--Length (inches) of the amount of overlap between shingles (Measured vertically).

INSIDE MITER--Trough corner extending toward roof.

PITCH--Slope of roof, measured in 12 foot lengths. Example, a 10/12 pitch indicates that the roof rises 10 feet for every 12 feet of run.

RAKE--The angled edge of a roof located at the end of a roof that extends past gable.

RIDGE--Intersection of any two roofing planes where water drains away from the intersection. Special shingles are applied to ridges.

RIDGE VENT--Opening at the point where roof decking normally intersects along the highest point on a roof where air is allowed to flow from the attic. A small cap covers this opening to prevent rain from entering. When these are long, they are normally known as continuous ridge vents.

SCREEN--Metallic or vinyl grid used to keep troughs free of leaves and other debris. Some screens are hinged.

SHINGLE--Uniform unit used in the coverage of a roofing deck. Shingles are manufactured from many materials including asphalt, fiberglass, wood shakes, tile and slate. Sizes normally vary from 12" X 36" to 12 1/4" X 36 1/4" with many sizes in between.

SHINGLE BUTT--The lower, exposed side of a shingle.

SIDE LAPP--Length (inches) of the amount of overlap between two horizontally adjoining shingles.

SOIL STACK--Corrosion resistant collar placed over vent pipes coming out of the roof deck.

SPIKE--A long nail used to attach the trough to the roof. Used in conjunction with ferrule. (See FERRULE).

SPLASHBLOCK--Special concrete form used to direct water from downspout and away from dwelling.

SQUARE--Roofing material required to roof 100 square feet of roof surface.

STACK--Also known as vent stack. A ventilation pipe coming out of the roofing deck.

STARTER STRIP--A continuous strip of asphalt roofing used as the first course, applied to hang over the eave.

UNDERLAYMENT--Any paper or felt composition used to separate the roofing deck from the shingles. Underlayments such as asphalt felt (tar paper) are common.

VALLEY--Intersection of two roofing planes where water drains at the intersection.

Notes

16. Plumbing

Plumbing is comprised of three interrelated systems:

- The water supply,
- The sewer system, and
- The (wet) vent system.

THE WATER SUPPLY

Unless you have opted for an individual water supply,(such as a well), you will tap into a public water supply. Unless you plan to live in your home many years (eight or more), the cost of drilling a well cannot be justified on cost alone. Remember, your water bill includes sewer service, too. You'll still pay for that (unless you opt for a septic tank or cesspool). If your local water supply is 80 psi or greater, you will need to install a pressure reduction valve at the main water service pipe. High water pressure can AND WILL take its toll on most pipes and fixtures in a short period of time.

Copper has been the material of choice by plumbers for water pipes for many years, but new materials available today have numerous advantages over copper. CPVC pipe is a plastic used for hot and cold supplies as well as its cousin PVC, which is only rated for cold water and sewer installations. CPVC offers several advantages over copper, including ease and speed of installation and some cost savings. Some people complain of a slight aftertaste in the water for the first year or so.

Polybutylene is a new semi flexible material that has taken the plumbing world by storm in recent years. It offers many advantages over copper and CPVC. It is totally inert and imparts no taste to the water. Its flexible nature allows it to be installed with a minimum of joints; which are the most expensive material in water systems. Polybutylene is also virtually freeze proof; a characteristic that can be appreciated by anyone who has had their pipes freeze in cold weather. It is very quiet in operation and reduces the effect of water hammer. Polybutylene joints are installed by one of two methods; crimping or grabber fittings. Crimp joints are the most cost effective, but the grabber joints are ridiculously easy to install. You simply insert the pipe into the fitting and push; that's it.

Polybutylene is by far the least expensive method of installing plumbing; but you may have trouble finding plumbers willing to work with it. Plumbers are slow to accept new innovations and may not pass the full savings of the installation to you. Many plumbers fail to realize the full cost savings available to them because they tend to install many more joints than are necessary. The flexibility of Polybutylene allows it to be wrapped and molded around obstructions; eliminating the need for many joints. Check with your local plumbing inspector for plumbers familiar with this new material. It is well worth the research.

THE SEWER SYSTEM

The sewer system collects all used water and waste and disposes of it properly. Unless you opted for a private sewer (septic tank), you will connect to the public sewer system. In most cases, building codes will not allow you to build a private sewer if a public one is available. When designing the house, keep in mind that sewer systems are powered by gravity. Hence, your lowest sink, toilet or drain must be higher than the sewer line. The sewer line must have at least 1/4" drop per foot in order to operate properly. This may mean raising your foundation in necessary to accommodate proper drainage.

With sewer pipe materials, technology has created a new problem. PVC pipe is now the most common material used because of its ease of installation; however, plastic pipes transmit the noise of flushing and water draining much more than the old cast iron pipes. Install cast iron sewer pipe if possible to reduce this noise. If you use PVC pipe, take care to position the pipes in walls that are isolated from bedrooms or you will constantly hear water noises.

THE (WET) VENT SYSTEM

Also known as the vertical wet venting system, this permits your plumbing system to "breathe". This system protects against siphonage and backpressure. All drains are attached to one of several vents in a network. These vents are seen as pipes sticking out the back side of your roof.

PLUMBING SUBS

Plumbers will normally bid and charge you a fixed price per fixture to install all three systems described above. You are likely to pay extra for jacuzzis, whirlpools and the like. The price of all fixtures (tubs, sinks, disposals) should be obtained from your sub. Fixtures above builder grade will usually cost more. If you use many custom fixtures or if your house plan positions bathrooms and the kitchen close together to conserve plumbing material, consider asking for a bid based on actual material and labor costs. This can save you a considerable amount of money. If you purchase fancy fixtures, have the plumber deduct the standard fixture charge for the ones you have purchased.

After you have received the estimate, ask for the cash price. Ask your plumber how many hours it will take him to do the job. You will pay the plumber 40 percent after passing the rough-in inspection and 60 percent after passing the final inspection. If your plumbing code does not require individual shut-off valves on each fixture, do not use them on each fixture if you want to save a little money.

BUILDING CODES

Most likely, your local building code will impose a number of requirements on your plumbing system. This helps protect you from paying for an inadequate system. Your local building code will cover plumbing materials and allowable uses of each, water supply and sewer capacities, joints, drainage, valves, testing, etc. READ THIS AND KNOW IT WELL. Although you should rely on your plumbing sub for advice, know your plumbing cold!

ABOUT SEPTIC TANKS

If you are considering going this route:

■ CONFIRM that septic tanks are permissible. Contact local inspectors and consult your local building code.

■ CONDUCT a percolation test to determine how quickly the soil absorbs water. Certain minimum standards will be imposed on your system. It is necessary to hire a qualified engineer to perform this test.

■ ESTIMATE your minimum septic tank capacity requirements.

■ REFER to your local building code for additional guidelines and requirements on a private sewer system.

PLUMBING: STEPS

ROUGH IN

PL1　DETERMINE type and quantity of plumbing fixtures (styles and colors). This includes:

■ Sinks (Kitchen, baths, utility, wet bar)
■ Bath tubs (consider one piece fiberglass units with built-in walls requiring no tile. If you go this route get a heavy gauge unit.)
■ Shower fixtures
■ Toilets and toilet seats
■ Water spigots (exterior)
■ Water heater
■ Garbage disposal
■ Septic tank
■ Sauna or steamroom
■ Water softener
■ Refrigerator ice maker
■ Any other plumbing related appliance

PL2　CONDUCT standard bidding process. Shop prices carefully, subs that bid on your job may vary greatly on their bid price, depending on what materials they use and their method of calculating their fee. This is where you will decide what kind of pipe to use.

PL3　ORDER special plumbing fixtures. This needs to be done well in advance because supply houses seldom stock large quantities of special fixtures. These will take some time to get in.

PL4　APPLY for water connection and sewer tap. You will probably have to pay a sewer tap fee.

PL5　INSTALL stub plumbing. This applies to instances where plumbing is cast in concrete foundations. This must be done after batter boards and strings are set and before concrete is scheduled.

PL6　Place all large plumbing fixtures such as large tubs, fiberglass shower stalls, and hot tubs before wall framing begins. These fixtures will not fit through normal stud or door openings.

PL7 MARK location of all plumbing fixtures including sinks, tubs, showers, toilets, outside spigots, wetbars, icemakers, utility tubs, washers, and water heater.

PL8 INSTALL rough-in plumbing. This involves the laying of hot and cold water lines, sewer and vent pipe. Pipe running along studs should run within holes drilled (not notched) in the studs. All pipe supports should be in place. Your plumber should use FHA straps to protect pipe from being pierced by drywall subs where ever cut-outs are made. FHA straps are metal plates designed to protect the pipe from puncturing by nails. The plumber will conduct a water pipe test using air pressure to insure against leaks.

PL9 INSTALL water meter and spigot. This has to be done early because masons will need a good source of water.

PL10 INSTALL sewer tap. Have the county sewer do this well before rough in plumbing is scheduled to insure that it is done on time.

PL11 SCHEDULE plumbing inspector.

PL12 INSTALL septic tank and line, if applicable.

PL13 CONDUCT rough-in plumbing inspection. Note: This is a VERY important step. No plumbing should be covered until your county inspector has issued and inspection certificate. Plan to go through the inspection with him so you will understand any problems and get a good interpretation of your sub's workmanship.

PL14 CORRECT any problems found during the inspection. Remember, you have the county's force behind you. You also have the specifications. Since your plumber has not been paid, you have plenty of leverage.

PL15 PAY plumbing sub for rough-in. Get him to sign a receipt or equivalent.

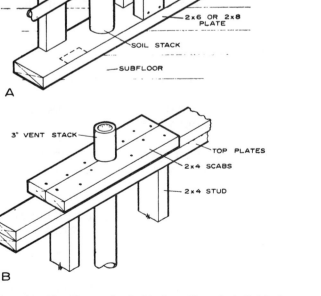

Figure 38. - Plumbing stacks: A, 4-inch cast-iron stack; B, 3-inch pipe for vent.

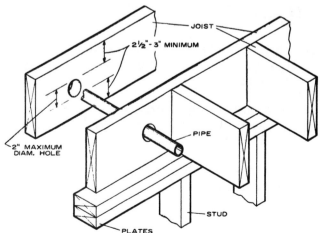

Figure 39. Drilled holes in joists.

FINISH

PL16 INSTALL finish plumbing. This involves installation of all fixtures selected earlier. Sinks, faucets, toilets and shower heads are installed .

PL17 TAP into water supply. This is where your plumbing gets a real test. You will have to open all of the faucets to allow air to bleed out of the system. The water will probably look dirty for a few minutes but don't be alarmed. The system needs to be flushed of excess debris and solvents.

PL18 CONDUCT finish plumbing inspection. Call your inspector several days prior to the inspection. Make every effort to perform this step with your local inspector to see what he is looking for so you will know exactly what to have your sub correct if anything is in error.

PL19 CORRECT any problems found during the final inspection.

PL20 PAY plumbing sub for finish. Have him sign an affidavit.

PL21 PAY plumber retainage.

PLUMBING: SAMPLE SPECIFICATIONS

■ Bid is to perform complete plumbing job for dwelling as described on attached drawings.

■ Bid is to include all material and labor including all fixtures listed on the attached drawings.

■ All materials and workmanship shall meet or exceed all requirements of the local plumbing code.

■ No plumbing, draining and venting to be covered, concealed or put into use until tested, inspected and approved by local inspectors.

■ All necessary licenses and permits to be obtained by plumber.

■ Plumbing inspection to be scheduled by plumber.

■ All plumbing lines shall be supported so as to insure proper alignment and prevent sagging.

■ Workman's compensation to be provided by plumber.

■ Floor drain pan to be installed under washer if located upstairs or for water heater in attic.

■ Install approved pressure reduction valve at water service pipe.

■ Stub in basement, plumbing for one sink, tub and toilet.

■ Drain pans to be installed in all tile shower bases.

■ Install construction water spigot on water meter with locked valve.

■ Install refrigerator ice-maker and drain.

■ Plumb water supply so as to completely eliminate water hammer.

■ Water pipes to be of adequate dimension to supply all necessary fixtures simultaneously.

SEWER SYSTEM

■ Hook into public water system

■ Install approved, listed and adequately sized backwater valve.

■ Horizontal drainage to be uniformly sloped not less than one inch in four feet toward point of disposal.

■ Drainage pipes to be adequate diameter to remove all water and waste in proper manner.

WET VENT SYSTEM

■ Vent piping shall extend through the roof flashing and terminate vertically not less than six (6) inches above the roof surface and not less than one (1) foot from any vertical surface, ten (10) feet from and three (3) feet above a window, door or air intake or less than ten (10) feet from a hot line.

■ All vent piping to be on rear side of roof without visibility from front.

■ All vent piping to conform to the local plumbing and building code.

PLUMBING: INSPECTION

ROUGH-IN

_____ All supplies and drains specified are present and of proper material. (PVC, CPVC, Copper, Cast iron, Polybutylene).

_____ Sewer tap done.

_____ No pipes pierced by nails.

_____ Hot water fixture on left of spigot.

_____ No evidence of any leaks. This is particularly critical at all joints, elbows and FHA straps.

_____ All plumbing lines inside of stud walls. Must allow for a flush wall.

_____ FHA straps used to protect pipes from nails where necessary.

_____ All cut-out framing done by plumber repaired so as to meet local building code.

_____ Tub centered in bath properly and is secured in place.

_____ Tub levelers are all in contact with tub. Tub does not move or rock.

_____ Toilet drain at least 12" to center from all adjacent walls to accommodate toilet fixture.

_____ Toilet drain at least 15" to center from tub to accommodate toilet fixture.

_____ Exterior spigot stub-outs high enough so that they will not be covered over by backfill. Minimum of 6".

_____ Ice-maker line for refrigerator in place. (1/4" copper pipe).

_____ Roof stacks have been properly flashed with galvanized sheet metal.

_____ All wall-hung sinks have metal bridge support located with center at proper level.

_____ Water heater firmly set and connected.

_____ Attic water heater has a floor drain pan.

_____ Water main shut-off valve works properly.

_____ Water lines to washer installed.

_____ Rough-in inspection approved and signed by local building inspector.

FINISH

_____ Tub faucets (hot and cold) operate properly. No drip. Drain operates well.

_____ Sink faucets (hot and cold) operate properly. No drip. Drain operates well.

_____ Toilets flush properly. Fill to proper line and action stops completely with no seepage.

_____ Kitchen sink faucets (hot and cold) operate properly. No drips. Drain operates well.

_____ Garbage disposal operates properly. Test as needed.

_____ Dishwasher operates properly (hot and cold). No leaks. Drain operates well. Run through one entire cycle.

_____ No scratches, chips, dents or other signs of damage on any appliances.

_____ No evidence of water hammer in entire system. Turn each faucet on and off very quickly and listen for a knock.

_____ All water supplies: Hot on the left and cold on the right.

_____ Turn on all sinks and flush all toilets at the same time and check for significant reduction in water flow. Some is to be expected.

_____ All exterior water spigots freeze proof and operate properly.

_____ All roof and exterior wall penetrations tested waterproof.

_____ Pipe holes in poured walls sealed with hydraulic cement.

Note: It is HIGHLY recommended that you obtain the assistance of your local inspector when performing your plumbing inspections. Your plumbing inspector is highly trained for such tests.

PLUMBING: GLOSSARY

AIR CHAMBER--Pipe appendage with trapped air added to a line to serve as a shock absorber to retard or eliminate air hammer.

BACKFLOW--The flow of water or other fluids or materials into the distributing source of potable water from any source other than its intended source.

BEND--Any change in direction of a line.

BIBB--Special cover used where lines are stubbed through an exterior wall.

BUILDING DRAIN--The common artery of the drainage system receiving the discharge from other drainage pipes inside the building conveying it to the sewer system outside the building.

BUILDING SEWER--That part of the drainage system extending from the building drain to a public or private sewer system.

BUILDING SUPPLY--The pipe carrying potable water from the water meter or other other water source to points of distribution throughout the building and lot.

CAP--Hardware used to terminate any line.

CLEAN-OUT--A sealed opening in a pipe which can be screwed off to unclog the line if necessary.

CPVC--Chlorinated Poly Vinyl Chloride. A flexible form of water line recently introduced. No soldering involved. Plastic.

ELBOW--A section of line which is used to change directions. Normally at right angles.

FIXTURE--Any end point in a plumbing system used as a source of potable water. Fixtures normally include sinks, tubs, showers, spigots, sprinkler systems, washer connections and other related items.

LINE--Any section of plumbing whether it is copper, PVC, CPVC or cast iron.

POTABLE WATER-- Water satisfactory for human consumption and domestic use, meeting the local health authority requirements.

PVC--Poly Vinyl Chloride. A form of plastic line used primarily for sewer and cold water supply.

SOIL STACK--A vent opening out to the roof which allows the plumbing system to equalize with external air pressure and to allow the sewer system to "breathe".

STOP VALVE--Shut-off valve allowing water to be cut-off at particular point in the system.

TRAP--A device providing a liquid seal which prevents the backflow of air without materially affecting the flow of sewage or waste water. "S" shaped drain traps are required in most building codes.

VENT SYSTEM--A pipe or network of pipes providing a flow of air to or from a drainage system to protect trap seals from siphonage or back pressure.

SEPTIC TANK--A receptacle used for storage of water, retained solids and digesting organic mater through bacteria and discharging liquids into the soil (subsurface, disposal fields or seepage pits), as permitted by local health authorities.

Notes

17. Heating - Ventilation - Air Conditioning

Heating, Ventilation and Air Conditioning is one of your big three trades in terms of cost and impact. Heating and air conditioning systems share the same ventilation and ductwork. Cold air, for air conditioning is heavier than hot air, hence it normally takes a higher speed fan. For this reason, see if you can get a two speed fan - one speed for cold air in the summer and one for hot air in the winter. The fan unit, normally a "squirrel cage" type, looks just like its name implies. If you have a two story home ,a large split level or a long ranch, consider a zoned (split) system. Zoned systems work on the premise that two smaller, dedicated systems operate more efficiently than one larger one. For large homes (2500 sq. ft. or more), this becomes a STRONG consideration. Heating is measured in BTU (British Thermal Units) while cooling is normally measured in tons (no relationship to 2000 lbs.). Normally, it takes 30,000 BTU of heat and one ton of AC for each 800 sq. feet of area.

The efficiency of heating and cooling systems varies widely. Select your systems carefully with this in mind. Contact your local power and gas companies for information regarding energy efficiency ratings used to compare systems on the market. When shopping for systems, check the efficiency rating, and be prepared to pay more for more efficient systems. Your power and gas company may also conduct, free of charge, an energy audit of the home you plan to build, and help suggest heating and cooling requirements. For efficiency reasons, your HVAC system should be located as close to the center of your home as possible, to minimize the length of ductwork required. Also, ductwork should have as few turns in it as possible. Air flow slows down every time a turn is made, decreasing the system's overall efficiency and effective output. Seven inch diameter round ductwork is fairly standard.

Air conditioning units operate by transferring heat from the air inside your home to the outside via a fluid called Freon. Several grades of Freon exist. Freon-16 is used in most AC units. Freon is measured by the pound (weight) - not volume. Make sure your system is fully charged before operating to avoid damage to the compressor.

OTHER CLIMATE CONTROL

At your option, your HVAC sub will also install other climate control devices. Electric air filters attach to your HVAC system air intake, eliminating up to 90 percent of airborne dust particles when cleaned regularly. Humidifiers, popular in dry areas, add moisture to the air. Special water supplies can be attached for automatic water dispensing. Dehumidifiers, popular in the Southeast and other humid areas, draw water out of the air. These units may be attached to a water drain for easy disposal of accumulated water.

HVAC SUBS

HVAC subs charge based on the size of your system (BTU and Tonnage). It is a tendency for subs to overestimate your heating and cooling requirements for three reasons:

■ More cost, more profit
■ Nobody ever complains about having too much heating and cooling!
■ Better safe than sorry

Have your HVAC sub submit a separate bid for the ductwork. Your HVAC sub will also install your natural gas line (if you use natural gas), and hook it up to the gas main. The gas company will install your gas meter. Shop carefully for your HVAC sub; a wrong decision here can cost you increased utility bills in the future.

DUCTWORK

The ductwork circulates conditioned air to specific points in the home and is split between air supplies which deliver conditioned air and air returns, which recirculate the air back to the HVAC system. Ducts will usually be constructed of sheetmetal or insulated fiberglass cut and shaped to fit at the site. Fiberglass ducts are rapidly gaining in popularity due to their ease of fabrication at the site. The duct itself is made of insulated fiberglass and can be cut with a knife. This extra insulation makes for a more efficient and quieter air system. Sheetmetal ducts are notorious for developing mysterious noises caused by expansion and air movement that may be difficult or impossible to eliminate.

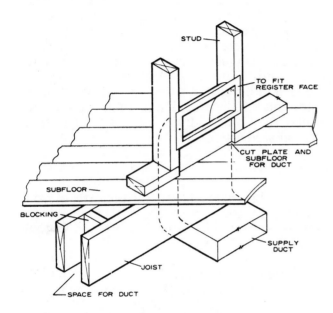

Figure 40. - Spaced joists for supply ducts.

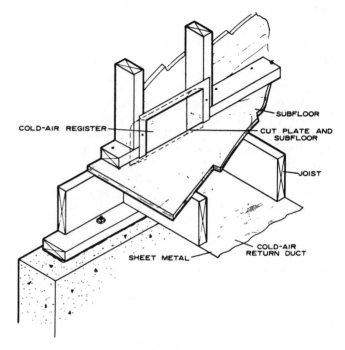

Figure 41. - Cold-air return.

REGISTER PLACEMENT
Proper placement of supplies and returns is critical to making your total HVAC system efficient and economical to run. Most registers are placed near doors or under windows. This type of placement puts the conditioned air near the sources of greatest heat loss and promotes more even distribution of air.

SUPPLIES should be located primarily along the exterior walls of the home. This is known as a radial duct system. Where possible, hot air supplies should be positioned directly below windows and along side exterior doors near the floor where possible since heat rises. Cold air supplies should be located up high in rooms since cool air falls.

RETURNS should be located low for warm air returns since the cooled air drops. Cool air returns should be located high to remove the air as it warms up. It is a good idea to have two supplies (high and low) and two returns (high and low) in critical rooms in order to handle either season. Your HVAC sub will be able to help you with the proper quantity and placement of your supplies and returns. As a general

rule, you will need at least one supply and one return for each room.

THERMOSTATS
Many elaborate thermostats are available on the market today. Many automatically shut heating systems down or set them at a low level during "off" hours. In any event, do not put the thermostat within six feet of any air supply register or facing one on an opposite wall. NEVER place a thermostat in a room which has a fireplace. Consider placing a thermostat in the hallway.

GET THE MOST OUT OF YOUR SUBS
If you are building in the Winter, subs will normally be used to working in a cold, unheated house. This means subs may not show up or when they do, they might spend a lot of time just getting warm by a fire somewhere. If you get your heating system approved early, normally by passing a quick test, you can turn it on before the house is finished. This will also save you some money when other builders may have gone out and rented a space heater. Drywall and paint will dry faster and subs will really appreciate the additional comfort.

HVAC: STEPS

HV1 CONDUCT energy audit of home to determine HVAC requirements. Your local gas and electric companies can be very helpful with this, sometimes providing computer printouts at little or no cost. You may only have to drop off a set of blueprints. Decide whether you will have a gas or electric dryer and determine where it will be situated.

HV2 SHOP for best combination of cost, size and efficiency in heating and cooling systems. The higher the efficiency, the higher the price.

HV3 CONDUCT bidding process on complete HVAC job. Have them give a separate bid for the ductwork.

HV4 FINALIZE HVAC design. Have a representative from the local gas company or an inspector look over your plan to make sure you aren't doing anything that could be a problem. Specify location of dryer exhaust vent.

HV5 ROUGH IN Heating and Air Conditioning system. Do not install any external fixtures such as compressors at this time. Compressors are a favorite theft item. If you frame floor joists 12" OC you may need to use floor registers.

HV6 INSPECT Heating and Air Conditioning rough-in. This should be done with your county inspector. Call him in advance.

HV7 CORRECT all deficiencies noted in the rough-in inspection.

HV8 PAY HVAC sub for rough-in work. Have him sign a receipt.

FINISH WORK

HV9 INSTALL Heating and Air Conditioning (finish work). This is where the thermostat and all other HVAC electrical work is hooked up. All registers are installed. The AC compressor is installed and charged with freon. Discuss the placement of the AC compressor location with the HVAC sub. He may supply the concrete pad if you do not have one.

HV10 INSPECT Heating and Air Conditioning final. Again, this should be done by your county inspector. Call him several days in advance so you can be present for the inspection. Some builders like to have their HVAC sub present, too. Make a note of any deficiencies. Be sure electrician has run service to the unit and that it is functional.

HV11 CORRECT all deficiencies noted in the final inspection.

HV12 CALL gas company to hook up gas lines.

HV13 DRAW gas line on plat diagram.

HV14 PAY HVAC sub for finish work. Get him to sign an affidavit.

HV15 PAY HVAC sub retainage after system is fully tested.

HVAC - SAMPLE SPECIFICATIONS

■ Bid is to install heating ventilation and air conditioning system (HVAC) as indicated on attached drawings.

■ Install zoned forced air gas (FAG) fired heating unit. With five (5) year warranty. One in basement and one in attic.

■ Install two AC compressors.

■ Install gas range hook-up.

■ Install 50-gallon gas water heater.

■ Install down draft line for island cooktop.

■ Connect 1 1/2" copper gas line to gas main, a minimum of twelve (12) inches under ground.

■ Install sloped PVC tubing for AC drainage.

■ Install gas grill line and grill unit on patio.

■ Install one fireplace log lighter and gas line.

■ Install gas line to porch and mailbox gas lights.

■ All work and materials to meet or exceed requirements of the local building code.

■ Provide and install all necessary air filters.

■ Opening to furnace large enough to permit removal and replacement.

■ All freon lines charged with freon.

HVAC - INSPECTION

ROUGH-IN

_____ All equipment UL approved with warranties on file.

_____ Heating and air units installed in place and well anchored.

_____ Heating and air units are proper make, model and size.

_____ Zoned systems have proper units in proper locations.

_____ Air compressors firmly anchored to footings.

_____ All ductwork installed according to specifications.

_____ Proper number of returns and supplies have been installed.

_____ All ductwork meets local building codes.

_____ All ductwork joints sealed tightly and smoothly with duct tape.

_____ No return ducts in baths or kitchen.

_____ Bathrooms vent to outside if specified.

_____ All ductwork in walls flush with walls, not to interfere with drywall.

_____ All ductwork in ceiling tied off so as not to interfere with drywall or paneling.

_____ Attic furnace has a floor drain pan, if in attic.

_____ All heat exhaust vents isolated from wood or roofing by at least one inch.

_____ No exhaust vents visible from the front of the home.

_____ Air conditioning condensate drain installed.

_____ All ductwork outlets framed properly to allow installation of vent covers.

_____ Vent hood or down draft installed in kitchen according to plan.

_____ HVAC rough-in inspection approved and signed by local building inspectors.

_____ Gas meter in place.

_____ Gas line connected to gas main.

FINISH

_____ Line to gas range hooked up. Range installed.

_____ Line to gas dryer hooked up. Dryer vent installed properly.

_____ Line to gas grill hooked up. Gas grill installed.

_____ HVAC electrical hook up completed.

_____ Thermostats operate properly and are installed near center of house. Thermostats located away from heat sources (fireplaces and registers) and doors and windows to provide accurate operation.

_____ Manual/Auto switch on thermostat works properly.

_____ A/C condensate pipe drains properly. This will be difficult or impossible to test in cold weather. The pipe should at least have a slight downward slope to it. You cannot test AC if temperature is below 68 degrees.

_____ Fan noise not excessive.

_____ Furnace, A/C and electronic air filters installed.

_____ Water line to humidifier installed and operating.

_____ Water line from dehumidifier installed and operating.

_____ All vent covers on all duct openings and installed in proper direction.

_____ All holes and openings to exterior sealed with exterior grade caulk.

_____ Downdraft vent for cooktop installed and operating.

_____ All HVAC equipment papers filed away. (warranty, maintenance, etc...)

_____ Strong air flow out of all air supplies.

_____ Air returns function properly.

HVAC: GLOSSARY

BTU--British Thermal Unit. A standard unit of hot or cold air output.

COMPRESSOR--Component of the central air-conditioning system which sits outside of dwelling.

DAMPER--A metal flap controlling the flow of conditioned air through ductwork.

ELBOW--Ductwork joint used to turn supply or return at any angle.

EXHAUST--Air saturated with carbon dioxide. The byproduct of natural gas combustion in a forced air gas system. This exhaust gas should be vented directly out the top of the roof. It is dangerous to breathe this gas.

FREON--A special liquid used by an air-conditioning compressor to move heat in or out of the dwelling. This fluid circulates in a closed system. This fluid is also used in refrigerators and freezers.

PLENUM--Chamber immediately outside of the HVAC unit where conditioned air feeds into all of the supplies.

REGISTER--Metal facing plate on wall where supply air is released into room and where air enters returns. Registers can be used to direct the flow of air.

RETURN--Ductwork leading back to the HVAC unit to be reconditioned.

SUPPLY--Ductwork leading from the HVAC unit to the registers.

TON--An industry standard measure to express a quantity of cold air produced by an air conditioning system.

18. Electrical

Electricians will perform all necessary wiring for all interior and exterior fixtures and appliances. Electricians will often charge a fixed price for providing and wiring each switch, outlet and fixture. Double switches will count as two switches and so on. You can save money by purchasing your own fixtures; the markup is incredible. Code will normally require at least one outlet per wall of each room and/or one for each specified number of running wall feet. The electrical code most often used and referred to is the National Electrical Code (NEC). But the one in your county should prevail. Your electrician needs to be a licensed master electrician in your county.

Your electrical sub will help you determine where switches and outlets should be placed. There are a few standards, such as minimum spacing of outlets and height of switches and outlets which should be indicated in your local building code. Consider multiple switches for a single light (such as at the top and bottom of stairs), floodlight switches in the master bedroom and photo-electric cells for driveway and porch lights. Make sure to install Ground Fault Interrupters (GFI's) in all bathroom outlets. GFI's are ultra sensitive circuit breakers that provide an extra margin of safety in areas where the risk of shock is high. Hobby shop and garage areas are also good sites for these devices.

Wire gauges used for circuits will vary. Electric stoves, ovens, refrigerators, washers, dryers and air compressors are each normally on separate circuits and use heavier gauges of wire. Your local building code will describe load limits on each circuit and the appropriate wire types and gauges to use.

After your home is dried in, the electrical sub will "rough in" all the electrical wiring, fuse box, and electrical HVAC connections. Make sure to have the HVAC and plumbing subcontractor complete their rough in before calling in the electrician. This will prevent any wiring from having to be cut or rerouted in order to make room for plumbing or ducting. It is much easier for the electrician to work around obstructions. If all ducting is in place, the electrician will also know precisely where the furnace will be located and will be able to wire for it properly.

Many house plans include a wiring diagram; but make sure to examine it carefully. These diagrams are usually done as an afterthought and are produced by the draftsman; not an electrician. It is best to walk through the house with the electrician and mark the locations of outlets and switches. A good electrician will make helpful suggestions on the location of fixtures.

ADDITIONAL WIRING

While the electrical wiring is being installed, consider installing other necessary wiring for appliances such as phones, alarm systems, TV antennas, doorbells, intercoms and computers. Ask your electrician for bids on installing this other wiring; although his price may be quite high. This type of wiring makes an excellent do it yourself project if you are eager to get involved. Installing this wiring before drywall is installed is easy and saves considerable installation expense later.

ELECTRICAL: STEPS

ROUGH IN

EL1 DETERMINE electrical requirements. Unless your set of blueprints includes a wiring diagram, you will want to decide where to place lighting fixtures, outlets and switches. Make sure no switches are blocked by an open door. You may want to consider furniture placement while you are doing this. Even if your blueprint has an electrical diagram, don't assume it is correct. Most of the time, they can be improved. You may wish to investigate the use of low voltage and fluorescent lighting.

EL2 SELECT electrical fixtures and appliances. Visit lighting distributor showrooms. Keep an eye out for attractive fluorescent lighting which will be a long-term energy saver (but cannot use a dimmer switch). Special orders should be done now due to delivery times. You should listen to the doorbell sound before purchasing.

EL3 DETERMINE if telephone company charges to wire home for modular phone system. Find out what they charge for and how much. Even if you plan wireless phones, install phone jacks in several locations.

EL4 CONDUCT standard bidding process. Have subs give a price to install each outlet, switch and fixture. He will normally charge extra for wiring the service panel and special work.

EL5 SCHEDULE to have phone company install modular phone wiring and jacks. If your phone company charges too much for this service, just have your electrician do the job or do it yourself. Also try to arrange to have your home phone number transferred to your new house later so that you can keep the same phone number.

EL6 APPLY for and obtain permission to hook-up your temporary pole to public power system. You may need to place a deposit. Perform this early so that hook-up will take place by the time the framers are on site.

EL7 INSTALL temporary electric pole.

EL8 PERFORM rough-in electrical. This involves installing wiring through holes in wall studs and above ceiling joists. All the wiring for light switches and outlets will be run to the location of the service panel. To insure that your outlets and switches are placed where you want them, mark their locations with chalk or a marker. Otherwise, the electrician will place them wherever he desires. It is important to get lights, switches and outlets exactly where you plan to place furniture and fittings.

EL9 INSTALL modular phone wiring and jacks and any other wiring desired. You may wish to run a few speaker and cable TV wires throughout the house. Staple wires to studs where wire enters the room so the sheetrock crew will notice them.

EL10 SCHEDULE electrical inspection. The electrician will usually do this.

EL11 INSPECT rough-in electrical. The county inspector will sign off the wiring. Remember that this MUST be done before final drywall is in place.

EL12 CORRECT any problems noted during rough-in inspection.

EL13 PAY electrical sub for rough-in work. Get a receipt or use canceled check as a receipt.

FINISH

EL14 PERFORM finish electrical work. This includes terminating all wiring appropriately (switches, outlets, etc...). Major electrical appliances such as refrigerators, washers, dryers, ovens, vent hoods, exhaust fans, garage door openers, door bells and other appliances are installed at this time. The air compressor will also be wired.

EL15 INSPECT finish electrical. Again, your county inspector must be scheduled. Have your electrician present so he will know just what to fix if there is a problem.

EL16 CORRECT electrical problems if any exist.

EL17 CALL Electrical utility to connect service.

EL18 PAY electrical sub, final. Have them sign an affidavit.

EL19 PAY electrical sub retainage after power is turned on and all switches and outlets are tested.

ELECTRICAL: SAMPLE SPECIFICATIONS

■ Bid is to perform complete electrical wiring per attached drawings.

■ Bid to include all supplies except lighting fixtures and appliances. Bid to include installation of light fixtures.

■ Bid to include furnishing temporary electric pole and temporary hook-up to power line.

■ Bid may include wiring and installation of the following items:

» 200 amp Service Panel with circuit breakers
» Light switches (1-way)
» Light switches (2-way)
» Furnace
» Dish washer
» Garbage disposal
» Microwave (built in)
» Door chime set
» Door chime button(s)
» Bath vent fans
» Jacuzzi pump motor
» Light fixtures (all rooms and exterior areas)
» Three-prong (grounded) interior outlets
» Three-prong (grounded) waterproof exterior outlets
» Electric range
» Double lamp exterior flood lights
» Electric garage door openers
» Washer
» Dryer
» Central Air Conditioning Compressor
» Climate control thermostats
» Electric hot water heater
» Hood fan
» Central vacuum system
» Intercom, radio units, speaker wires
» Co-axial cable TV lines & computer cables
» Sump pump
» Electronic security system with sensors
» Ceiling fans
» Whole house attic fan
» Automatic closet switches
» Humidifier/de-humidifier
» Time-controlled heat lamps
» Thermostatic Roof vents
» Phone lines
» Time controlled sprinkler system

■ All work and materials to meet or exceed all requirements of the National Electrical Code unless otherwise specified.

■ Natural earth ground to be used.

ELECTRICAL - INSPECTION

ROUGH-IN

_____ All outlets and switches placed as indicated and are unobstructed by doors.

_____ All specified outlets, switches and fixtures are in place and work as intended. Refer to your specifications and blueprints.

_____ All lines are grounded.

_____ All visible splices have approved splice cap securely fastened.

_____ Bath ventilator fans installed

_____ Attic power ventilators installed

FINISH

_____ All outlets and fixture wires measure 117 V with circuit tester or voltmeter. Test between the ground and each socket.

_____ Dimmer switches installed where specified

_____ Air conditioner cut-off switch is installed at A/C unit.

_____ No scratches, dents or other damage to electrical appliances and fixtures.

_____ Furnace completely connected and operational.

_____ All bath vents connected and operational. Should be on separate circuit from light switch.

_____ Electric range connected and operational.

_____ Electric pilot light starter on gas range connected and operational.

_____ Range hood, down draft and light connected and operational.

____ Built-in microwave connected and operational.

____ Garbage disposal connected and operational.

____ Trash compactor connected and operational.

____ Dishwasher connected and operational.

____ Refrigerator outlet operational.

____ Door chimes (all sets), connected and operational.

____ Power/thermostatic ventilators on roof connected and operational.

____ Washer and Dryer outlet operational.

____ Garage opener installed and operational.

____ Electronic security system connected and operational. All sensors need to be tested one at a time.

____ Exterior lighting fixtures installed and operational

____ All recessed lighting, ceiling lighting, heat lamps and other lighting fixtures connected as specified and are operational.

____ All intercom/radio units connected and operational.

____ Jacuzzi pump motor connected and operational. For safety reasons, switch should not be reachable from the tub.

____ Central vacuum system connected and operational.

____ All other specified appliances and fixtures connected and operational.

____ All switch plate and outlet covers are installed as specified.

____ Finish electrical inspection approved and signed by local building inspector.

ELECTRICAL: GLOSSARY

AMPERAGE--The amount of current flow in a wire. Similar to the amount of water flowing in a pipe.

CAN--Recessed lighting fixture

CIRCUIT BREAKERS--A device to ensure that current overloads to not occur. A circuit breaker "breaks the circuit" when a dangerous overload or short circuit occurs.

CONDUIT--Metal pipe used to run wiring through when extra protection of wiring is needed or wiring is to be exposed.

EER--Energy Efficiency Rating - A national rating required to be displayed on appliances measuring their efficient use of electrical power.

GEM BOX--A metal box installed in electrical rough-in that holds outlets, receptacles and other electrical units.

GFI--Ground Fault Interrupter - An extra sensitive circuit breaker usually installed in outlets in bathrooms and exterior locations to provide additional protection against shock. Required now by most building codes.

HOT--A wire carrying current.

SERVICE PANEL--Junction where main electrical service to the home is split among the many circuits internal to the home. Circuit breakers should exist on each internal circuit.

THINWALL--Thin flexible conduit used between outlet boxes.

3 WAY SWITCH--Type of switch that allows control of a lighting fixture from two switch locations such as both ends of a hallway.

VOLTAGE--The "force" of electrical potential. Similar to the pressure of water in a pipe.

WATTAGE--The product of the amperage times the voltage. A good indicator of the amount of electrical power needed to run the particular appliance. The higher the wattage, the more electricity used per hour.

Notes

19. Masonry and Stucco

Masonry is the trade involved with the application of concrete blocks, clay or tile bricks and stone. Masonry is a true art; watch a bricklayer some time and you will see why. A skilled bricklayer possesses a strong combination of good workmanship and efficiency - key ingredients to producing a strong, well built brick structure on schedule.

CONCRETE BLOCKS

Normally used below ground as an alternative to a poured foundation. Although a poured foundation is three times stronger, concrete blocks are somewhat less expensive, while being of adequate strength. The standard size of a concrete block is 7 5/8" X 7 5/8" X 15 5/8" and are normally laid with 3/8" flush mortar joints. Concrete blocks are applied on concrete footings.

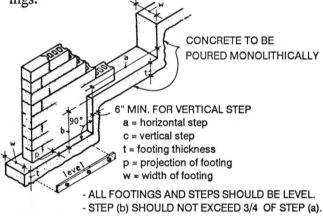

CONCRETE TO BE POURED MONOLITHICALLY

6" MIN. FOR VERTICAL STEP
a = horizontal step
c = vertical step
t = footing thickness
p = projection of footing
w = width of footing

- ALL FOOTINGS AND STEPS SHOULD BE LEVEL.
- STEP (b) SHOULD NOT EXCEED 3/4 OF STEP (a).

Figure 42.

CLAY BRICKS

Often called brick veneer, brick facing is normally comprised of brick 3 5/8" X 7 5/8" X 2 1/4" set in a 3/8" mortar joint. For estimating purposes, brick veneer, laid in running bond (a typical pattern), requires seven bricks per square feet of wall. This does not apply for such

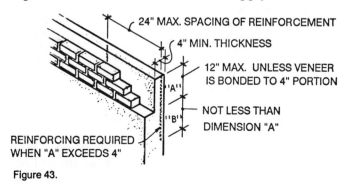

24" MAX. SPACING OF REINFORCEMENT

4" MIN. THICKNESS

12" MAX. UNLESS VENEER IS BONDED TO 4" PORTION

"A"

NOT LESS THAN DIMENSION "A"

"B"

REINFORCING REQUIRED WHEN "A" EXCEEDS 4"

Figure 43.

items as chimneys or fancy brickwork. Your mason charges for laying units of 1000 bricks (one skid). Additional expenses may be charged for multi-story or fancy brickwork where scaffolding and or additional skill and labor is required.

STUCCO

Once popular and widely used throughout Western Europe, stucco has come back in style to bring back that old flair. Stucco is really just mortar applied over a metal lath or screen used for support and adhesion.

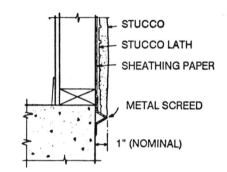

STUCCO
STUCCO LATH
SHEATHING PAPER
METAL SCREED
1" (NOMINAL)

NOTE: Base screed shown on concrete slab construction without sheathing screeds or other foundations or on sheathing are similar.

Figure 44. - Typical stucco base screed.

DURABILITY. Some builders complain of maintenance problems (cracking) of traditional stucco after ten years or so of service. IF stucco is applied properly with proper expansion joints, it should provide decades of sound performance. Expansion joints should be placed approximately every 300 square feet. Some stucco cracking should be expected.

INSULATION. Stucco is not the best insulator. As such you should go with heavier batt insulation in your exterior walls. Higher grade sheathing is also in order.

NEW IMPROVED. Several brands of artificial stucco are on the market. Although the boast of higher insulating qualities and lower maintenance, they are also more expensive. Some say it pays off in use of cheaper insulation and lower utilities. It is also easier to do fancier trim work like lintels, quoins and other work because it uses styrofoam cutouts as a base.

COSTS. Stucco is charged by the installed square yard. Normally, area is not deducted for doors and windows except for the garage door. Rationale is that it is at least as much trouble to work around an opening as it is to cover the area. Extras will probably include :

- Smooth finish
- Decorative corner work (quoins)
- Decorative window and door work (keystones & bands)
- Corner bead - all edges

Stucco can be applied rough, smooth or anywhere in between. Subs will charge you more for a smooth finish because it takes longer and imperfections show up easier. If you use a dyed stucco, it may not require painting.

FOR THE BEST STUCCO JOB

- Don't attempt to stucco over a foil or polyethylene backed sheathing due to poor adherence. Use either styrofoam (5/8" to 2") or exterior gypsum (1/2"). Check to see what sheathing your sub recommends.

- Avoid mixing pigments with stucco mortar. This can lead to uneven drying and coloration. Paint the stucco after it is dry.

- To avoid cracking, be liberal with expansion joints. They are hardly noticeable and do not detract from the overall appearance of the job.

- Don't attempt to apply stucco if there is ANY chance of rain in the forecast.

ABOUT FIREPLACES

If you go with a custom masonry fireplace, the critical thing to get is a proper draft. If your fireplace doesn't "breathe" properly, you will always have smoke in your home during its use. If it "breathes" too well, you may not get too much effect from the fireplace. This is not a big problem with pre-fabricated fireplace units since they are pre-designed to operate properly. Below are a few options you may want to consider regardless of what type fireplace you have:

Fresh air vent--This allows the fire to get its combustion air from outside your home so that it doesn't rob all of your warm air.

Ash dump--Makes fireplace cleaning easier.

Log lighter--Easier log starting.

Raised hearth--Easier access to working on the fire.

Blower--May provide more room heating.

Heat Recirculator--May also be able to provide more heating.

One important safety issue is the flue. Avoid burning pine whenever possible especially if you are using a prefab fireplace. The resin buildup in the flue can cause problems later (fire hazard). If you plan to have a stucco home, consider a prefab fireplace. A masonry fireplace can look out of place unless covered over with stucco; which isn't very cost effective!

MASONRY: STEPS

MA1 DETERMINE brick pattern, color and coverage.

MA2 DETERMINE stucco pattern, color and coverage.

MA3 PERFORM standard bidding process. Contact material suppliers, local brick companies, brick labor unions and other subs for the names of good masons. Do not select a mason until you have seen his work. Ask the suppliers if there is a penalty for returning unused bricks.

MA4 PERFORM standard bidding process for stucco subs. You really need to see their work, too.

MA5 APPLY flashing above all windows and doors.

MA6 LAY brick veneer exterior, chimney, fireplace and hearth. This is the time to have the mason throw in a brick mailbox if you want one.

MA7 FINISH bricks and mortar.

MA8 INSPECT brickwork. Refer to specification, local code and checklist on following page.

MA9 CORRECT any brickwork requiring attention.

MA10 PAY mason. If you have extra bricks which either can't be returned or are not worth returning, turn them into a brick walkway or patio later.

MA11 CLEAN up excess bricks and dried mortar. Spread any extra sand in the driveway area. Masons charge too much by the hour to have them do this work.

MA12 Install base for decorative stucco. This includes keys, bands, and other trimwork. Use exterior grade lumber.

MA13 PREPARE stucco test area. This should be at least a 3' by 6' sheet of plywood covered with test stucco to make sure you agree with the actual color and texture. Make sub finish mailbox if desired. If you wish to protect the stoops from stucco, cover them with sand.

MA14 APPLY stucco lath and decorative formwork to sheathing.

MA15 APPLY first coat of stucco. The color of the first coat does not matter.

MA16 APPLY second coat of stucco. You need to wait a few days for the first coat to dry completely. Place a thick mat of straw at base of house to prevent mud from staining stucco. While the scaffolding is set up, have the painter work on cornice and second story windows.

MA17 INSPECT stucco work.

MA18 CORRECT any stucco work requiring attention.

MA19 PAY stucco subs.

MA20 PAY mason's retainage.

MA21 PAY stucco sub's retainage.

MASONRY: SAMPLE SPECIFICATIONS

■ Brick veneer coverage to be as indicated by attached drawings.

■ All brick and mortar to be furnished by builder.

■ All necessary scaffolding to be supplied and erected by mason.

■ Brick entrance steps to be laid as indicated in attached drawings.

■ Brick patio to be laid as indicated in attached drawings.

■ All excess bricks to remain stacked on site.

■ Wrought iron railings to be installed on entrance steps as indicated in attached drawings. Wrought iron to be anchored into surface with sufficient bolts.

■ Bricks are to be moistened before laying to provide superior bonding.

■ Window, door, facia and other special brick patterns are to be done as indicated in attached detail drawings.

■ Moistened brick is to be laid in running bond with 3/8" mortar joints

■ Mortar is to be medium gray in color, Type "C" mortar is to be used throughout job.

■ Mortar joints are to be tooled concave.

■ Steel lintels are to be installed above all door and window openings.

■ Masonry walls are to be reasonably free of mortar stains as determined by builder. Mortar stains are to be removed with a solution of one part commercial muriatic acid and nine parts water to sections of 15 square feet of moistened brick and then washed with water immediately. Door and window frames are to be protected from the cleaning solution.

■ Flashing to be installed at the head and sill of all window openings.

■ Flashing to be installed above foundation sill and below all masonry work.

■ Bid to include all necessary touch-up work.

STUCCO: SAMPLE SPECIFICATIONS

■ Bid is to provide all material, labor and tools to install lath and stucco per attached drawings.

■ Stucco to be finished smooth, sand texture. (specify color)

■ All trim areas to be 2" wide unless otherwise specified.

■ Quoins to have beveled edges; alternating 12" and 8" wide.

■ Two coats of stucco to be applied; pigment in second coat.

■ Stucco work to come with a twenty year warranty for material and labor against defects, chipping and cracks.

■ Expansion joints to be as indicated on blueprint or as agreed upon.

MASONRY AND STUCCO: INSPECTION

____ Brick style and color are as specified.

____ Mortar color and tooling are as specified.

____ All wall sections plumb, checking with plumb bob and string. No course should be out of line more than 1/8". Overhang distance between top course and bottom course should not exceed 1/8".

____ Flashing installed above all windows, exterior doors and foundation sill.

____ No significant mortar stains visible from up close.

____ Stoops and patio steps drain properly.

____ No loose brick.

____ All quoins smooth, even and as specified.

____ Touch-up brickwork finished.

____ All wrought iron railing is complete per drawings and is sturdy.

STUCCO

____ Stucco coverage, colors and materials are as specified.

____ Stucco is finished as specified.

____ Stucco finish is even with no adverse patterns in surface. Surface level with no dips or bumps. Sight down surface from corners.

____ Stucco is applied to proper depth.

____ All decorative stucco work such as quoins, keys and special trim are in place, complete as specified and well finished with sharp edges and corners.

____ Adequate number of expansion joints. (one approximately every 300 square feet.) Expansion symmetrical between window openings.

____ Openings for exterior outlets and fixtures not covered up.

____ No large cracks.

MASONRY AND STUCCO ■ 159

MASONRY & STUCCO: GLOSSARY

ACID--Clearing agent used to clean brick.

BAND--Decorative trim around windows and doors and horizontal relief; normally 2X4 or 2X6.

BEAD--Any corner or edge that must be finished off with stucco.

BRICK VENEER--Exterior wall covered with brick.

CORBEL--Extending a course or courses of bricks beyond the face of a wall. No course should extend more than two inches beyond the course below it. Total corbeling projection should not exceed wall thickness.

COURSE--A horizontally laid set of bricks. A 32 course wall with 3/8" mortar joints stands 7 ' tall.

CUBE--A standard ordering unit for masonry block units (6x6x8).

JOINT--Mortar in between bricks or blocks.

KEY--Fancy decorative lintel above window made of brick normally place on various angles for a flared effect. Also known as a keystone.

LATH--A grid of some sort (normally metal or fiberglass) applied to exterior sheathing as a base for stucco.

LINTEL--A structural member placed above doors and window openings to support the weight of the bricks above the opening.

QUOIN--Fancy edging on outside corners made of brick veneer or stucco.

ROW LOCK--Intersecting bricks which overlap on outside corners.

SOLIDS--Solid bricks used for fireplace hearths, stoops, patios, or driveways.

TRIG --A string support for guide lines to prevent sagging and wind disturbances on long expanses of wall.

WATER LINE--Decorative relief line around foundation approximately three feet from the ground.

WEEP HOLE--Small gap in brick wall, normally on garage, that allows water to drain.

WOOD MOLD--A brick making process where bricks are actually molded instead of extruded. Fancy shapes are the result of this process.

Notes

20. Siding and Cornice

The siding and cornice sub will handle all siding application and cornice work (soffit, facia, frieze and eave vents). This sub can also install windows and exterior doors, if needed.

SIDING

Siding is the most economical exterior covering. The most popular materials include cedar, vinyl, aluminum and masonite. There are advantages and disadvantages to each:

Type	Advantages	Disadvantages
Cedar	Natural appearance Moderate cost	Higher maintenance May raise fire insurance premium
Texture 111	Inexpensive	
Masonite	Inexpensive	High maintenance Requires painting
Vinyl	Washes easily Low maintenance	Expensive
Aluminum	Washes easily Low maintenance	Expensive

WOOD SIDING provides some of the greatest variety in styles and textures as well as cost effectiveness. It can be installed horizontally, vertically, diagonally, or a combination to provide a creative and varied exterior.

Some of the wood types available include pine, cedar, cypress, and redwood. The full range of types and styles exceed the scope of this book. Check with your local supplier for the styles available in your area.

Regardless of the type of wood siding you choose, the important properties required for quality are freedom from warpage, knots and imperfections. Wood siding is prone to shrinkage after installation due to drying. If painted right after installation, lap siding will continue to shrink and leave an unpainted strip of wood where the pieces overlap. Try to delay painting until the siding has an opportunity to stabilize.

Drying will also cause several boards to warp and crack as they shrink. Always retain a portion of your siding sub's payment so that you can get him to return later to replace these defective pieces.

The introduction of several plywood siding types such as Texture 111 has simplified siding installation. These plywood sheets simulate several styles of wood siding such as board and batten, but are much easier to install and do not shrink. Labor costs to install are also much lower than traditional wood siding.

Wood siding is prone to leakage around windows and doors; especially on contemporary designs that have many angled walls. Therefore it is important that your siding sub or painter caulk all butt joints. Specify a tinted caulk that matches the final color of the siding.

MASONITE SIDING has many of the aesthetic qualities of wood with the added advantage of more uniform and stable material. Masonite is usually free of imperfections and will not shrink. It also has the added advantage of being preprimed, which may eliminate one paint coat. Most traditional styles that use siding have gone to masonite; however the introduction of new textures that simulate wood have made masonite more popular for rustic and contemporary designs as well.

The nail holes in masonite are much more visible than in wood, so watch your sub carefully during installation. Specify that the nails to be used are galvanized and countersunk.

VINYL AND ALUMINUM SIDING is usually installed by a subcontractor that supplies labor and materials for the job. Although expensive, these siding types will last for many years with little or no maintenance. Many come with a 20 year unconditional guarantee.

CORNICE

The cornice is the finishing applied to the overhang of the roof called the eave. The overhang of the roof serves a decorative function as well as protecting the siding from water stains and leakage. Generally the greater the overhang, the more expensive the look of the house.

The cornice will installed by your siding or trim sub. If you plan to use fancy trim around your

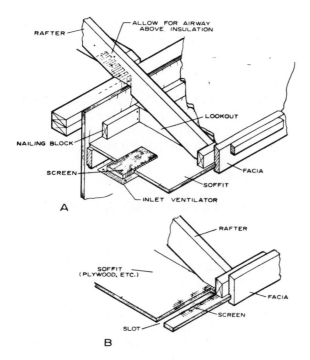

Figure 45. - Inlet ventilators: A, Small insert ventilator; B, slot ventilator.

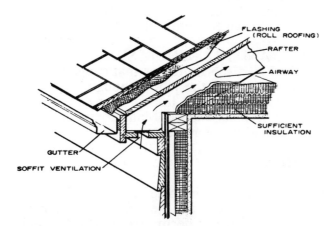

Figure 47. - Eave protection for snow and ice dams. Lay smooth surface 45-pound roll roofing on roof sheathing over the eaves extending upward well above the inside line of the wall.

cornice such as dentil molding, you may want to trust the job to your trim sub. He is more used to working with decorative trim.

The two types of cornices most commonly used are the open cornice and the box cornice. The open cornice is the simplest type and is more appropriate for contemporary styles. The open cornice looks just as it sounds. The rafter overhang is left unfinished and open and requires care in the choice of roof sheathing, since it will be visible from below. The box cornice has the overhang boxed in with plywood and is finished off with a variety of trim types and styles. This style is more commonly used on traditional styling and has the added advantage of providing a place to install eave ventilation; which will improve the energy efficiency of the house.

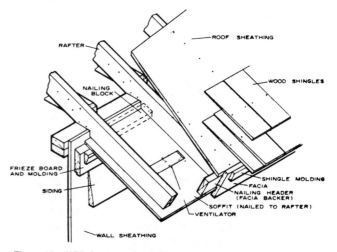

Figure 46. - Wide bow cornice (without lookouts).

The style and quality of the cornice can have a big effect on the appearance of your house, so take care in the design. If you use a draftsman or architect, ask him to provide a cornice detail drawing so that its design will be clear to the cornice sub. This will also help you in estimating materials for the job.

Cornice work is installed before stucco; so if you have that type of finish, be sure that the cornice sub builds the cornice out (usually with 1x6 or 2x6) so that the cornice sits out beyond the stucco. Once installed, the cornice should be primed with exterior paint as soon as possible. For convenience you may wish to have your painters prime the cornice material prior to installation.

SIDING AND CORNICE: STEPS

SC1   SELECT siding material, color, style and coverage.

SC2   CONDUCT standard bidding process. Most often the siding and cornice will be bid as a package deal.

SC3   INSTALL windows. Have wood shims available to adjust windows properly. Inspect window positioning. Windows should be plumb and secure.

SC4   INSTALL flashing above windows, exterior door openings and around bottom perimeter of house.

SC5   INSTALL siding.

SC6   INSTALL siding trim around corners, windows, and doors.

SC7   CAULK all areas where siding butts against trim or another piece of siding.

SC8   INSPECT siding. Refer to related checklist and specifications.

SC9   PAY sub for siding work. Make sure to retain a portion of payment for call backs.

SC10   INSTALL cornice. This involves installing the cornice, facia, frieze, soffit and eave vents. If you have a traditional front, you may want to use a dentil molding. Stucco fronts normally require an exterior grade of crown molding ranging from 3" to 6" thick.

SC11   INSPECT cornice work.

SC12   CORRECT any problems noted with cornice work.

SC13   PAY for cornice work.

SC14   CORRECT any problems noted with siding after it has had time to shrink.

SC15   PAY siding retainage.

SC16   PAY cornice retainage.

SIDING AND CORNICE: SPECIFICATIONS

SIDING

- Exterior galvanized siding nails to be used exclusively.

- All nails are to be flush or counter sunk.

- All laps of siding to be parallel.

- Joints to be staggered between courses.

- All work and materials are to conform to the local building code.

- All siding edges to be terminated in a finished manner and caulked.

- All openings and trim caulked and flashed.

CORNICE

- All soffit and facia joints to be trimmed smooth and fit tight.

- All facia to be straight and true.

- Soffit vents to be installed.

- All soffit, facia, frieze, and trim material to be paint grade fir or equivalent.

Notes

21. Insulation and Soundproofing

With the cost of most energy forms on the rise, insulation is gaining importance. The insulating property of a material is measured as an R Factor. The thickness of a material has little to do with its R rating. Don't buy thick insulation, buy high R-rated insulation. Insulation normally comes as fiberglass or cellulose. Either can be blown into most void areas, while fiberglass also comes in batts.

Insulation requirements vary across the country. Contact your local power company and they will assist you with recommended R ratings for ceilings, floors, walls, attics, etc... Observe that their recommendations may exceed your minimum local code requirements. Don't worry. You can pay for insulation now or pay for energy later. R factor requirements are not necessarily the same for all walls. Exposures which receive heavy sunlight or wind should be more heavily insulated. This is another job that you can do yourself. If you choose to insulate yourself, ask a supplier for a copy of the manufacturer's recommended installation instructions. Use gloves and an air filter over your nose.

SHEATHING
Sheathing is a light weight, rigid panel nailed to the outer side of the exterior walls. Several different types of sheathing are now available; each with different properties and costs. Foil covered sheathing has several advantages including higher insulation values, a finish more resistant to abrasion during installation, and greater airtightness. Its also one of the most expensive types. Sheathing can add considerably to the insulating value of your house, so research your insulation needs carefully before deciding which type to use.

FIBERGLASS BATS
These are rolled sections of fiberglass bonded to a sheet of heavy duty kraft paper. Fiberglass, as well as most other forms of insulation, lose their effectiveness when wet, moist, compressed or otherwise not installed in accordance with the manufacturer's instructions. Bats should fit snugly between studs and joists with no gaps. Many subs staple them in place.

BLOWN MATERIALS
Blown in materials are very popular as attic insulation because of ease of installation and the ability to fill in around odd shaped areas and framing. It is usually not suitable as wall insulation however, because of its tendency to settle. Both fiberglass, rock wool and cellulose can be blown in (as a loose material). This type of insulation is normally applied with depths ranging from six to twelve inches. Cellulose is the best bargain, but is susceptible to rodents and moisture. Beware of cellulose that does not conform to local fire treatment standards. Rock wool is a good compromise between quality and cost. Blown insulation makes a good do-it-yourself project. You can usually rent the blowers from the insulation supplier.

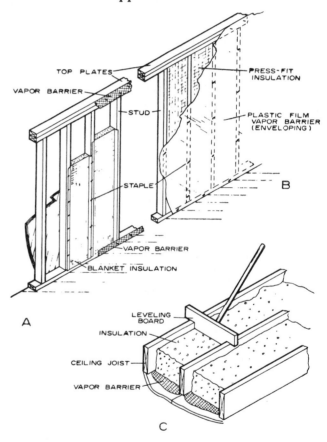

Figure 48. - Application of insulation: A, Wall section with blanket type; B, wall section with "press-fit" insulation; C, ceiling with full insulation.

SOUNDPROOFING
This is a refined touch. Consider soundproofing bathrooms and the HVAC area to quiet down the system. Rooms where you plan to have stereo systems are also good candidates. Soundproofing material comes in 4x8' sheets and can be applied between the sheetrock and studs.

INSULATION SUBS

Insulation subs should give you a price for material and labor. Insulation will normally be charged by the square foot for batts and by the cubic foot for blown in insulation. Investigate whether you qualify for any energy credit.

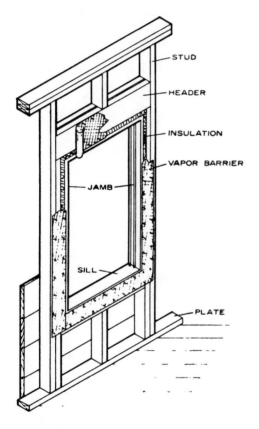

Figure 49. - Precautions to be taken in insulating around openings.

INSULATION: STEPS

IN1   DETERMINE insulation requirements. Your local energy company can help give you guidelines and suggestions.

IN2   PERFORM standard bidding process.

IN3   INSTALL wall insulation.

IN4   INSTALL soundproofing. This is a nice touch to deaden noise. Use in bathroom walls to deaden sound.

IN5   INSTALL floor insulation. Floor insulation will be installed on crawl space and basement foundations. Metal wires cut to the length of joist spacing are used to hold the insulation in place.

IN6   INSTALL attic insulation. This normally consists of batts or blown in insulation. If blown in, attic insulation will be installed after drywall has been nailed into place on the ceiling.

IN7   INSPECT insulation. A kraft paper or poly vapor barrier should be installed on the warm side of the insulation. All areas around plumbing, electrical fixtures, doors and windows should be stuffed with insulation to prevent air infiltration.

IN8   CORRECT insulation work as needed.

IN9   PAY insulation sub and have him sign an affidavit.

IN10   PAY insulation sub retainage.

INSULATION: SAMPLE SPECIFICATIONS

■ Bid is to provide and install insulation as indicated below:

Location	R Factor	Type
Walls	R13	Fiberglass batts
Ceiling	R19	Blown rock wool

■ Bid is to include labor and materials for wall and ceiling insulation.

■ All insulation is to include a vapor barrier on the warm side of the insulation, free of rips and reasonably sealed against air infiltration.

■ All joints and gaps around door frames, window frames and electrical outlets to be packed with fiberglass insulation.

■ All insulation used will be of thickness sufficient to meet R values specified after insulation has settled.

INSULATION: INSPECTION

_____ All specified insulation has been installed per specifications. Check labels on insulation.

_____ All insulation is installed tightly with no air gaps.

_____ No punctures have been made in vapor barrier.

_____ Insulation adheres firmly to all adjoining surfaces.

_____ No eave vents have been covered by insulation.

_____ Vapor barrier installed on warm side of wall.

_____ Ductwork insulated in basement.

_____ Plumbing in basement insulated.

_____ All gaps around doors and windows stuffed with insulation.

_____ Gap between wall and floor caulked completely.

_____ All openings to the outside for plumbing, wiring, and gas lines sealed with spray foam insulation.

_____ All gaps in siding around windows and corners caulked thoroughly.

Notes

22. Drywall

Drywall is composed of gypsum sandwiched between two layers of heavy gauge paper. This material, normally used for interior walls, comes in 4' x 8', 4' x 10, and 4' x 12' sizes. When planning your home, it may be helpful to plan around these dimensions to some extent to minimize waste.

Drywall comes in two common thicknesses: 3/8" and 5/8". 5/8" is the thickness most often used because it does not warp as easily, especially where studs are 24" on center (O.C.). Drywall sheets have the long edges tapered and the short edges full thickness. This is to allow room for the drywall mud and tape. Sheets should be installed so that the full edges butt against studs. After drywall is applied to studs, the joints between sheets of drywall are smoothed using special drywall tape. Moisture affects drywall adversely, so always store it in a dry place until used.

Drywall installation normally involves the following steps:

- Apply drywall to studs with either nails or screws, and glue.
- Apply fill coat of joint compound and drywall tape.
- Sand all joints smooth (if necessary)
- Apply joint compound (second coat)
- Sand all joints smooth (if necessary)
- Apply finish coat of joint compound (third coat)
- Sand all joints smooth

Application should not be done in extreme heat or cold. Ceilings are treated just as walls, unless you decide to stipple them. Stipple is a coarse-textured compound which eliminates the need for repeated sanding and joint compound application. This is a time and money saver.

DRYWALL SUBS

Your drywall sub is a KEY player in the final, visible quality of most surfaces in the interior. Drywall sub-contractors normally bid and charge for drywall work by the square foot, although they install it by the sheet. Of all your sub-contractors, quality here is important. Drywall is one of the most visible parts of your home. Get the best drywall sub you can afford. Below are a few signs of a good drywall sub:

- Finishing coats are very thin and feathered out
- No need to sand first two coats because they are so smooth
- Cutting pieces in place to assure proper fit.

Warning: It may be necessary for your drywall sub to use stilts to reach ceilings and high walls. Stilts are dangerous, but many times virtually unavoidable. Some states will not award Workman's Compensation to subs injured while using stilts. If this is true in your state, have your sub sign a waiver of liability to protect you from accidents. Scaffolding should be used for stairway work. If stilts are used, the floor should be cleared of scrap wood and trash to avoid tripping.

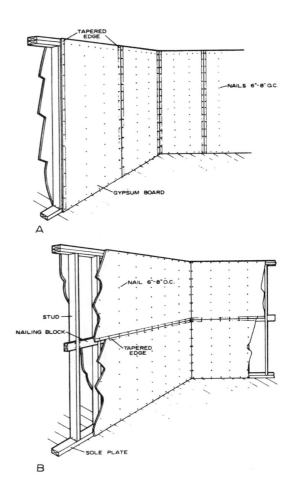

Figure 50. - Application of gypsum board finish: A, vertical application; B, horizontal application.

DRYWALL: STEPS

DR1 PERFORM drywall bidding process (refer to standard bidding process). Refer to drywall contract specifications for ideas. You may wish to ask painters for names of good drywall subs; they know good finishing work - they paint over it for a living!

DR2 ORDER and receive drywall materials if not supplied by drywall sub.

NOTE: Before drywall is installed, mark the location of all studs on the floor with a builder's pencil. This will make it easier for the trim sub to locate studs when nailing trim to the wall.

DR3 HANG drywall on all walls. Outside corners must be protected with metal edging. Inside corners must also be taped.

DR4 FINISH drywall. Drywall must be finished in steps as outlined below:

■ Spackle all nail dimples.
■ Sand nail dimples smooth.
■ Apply tape to smoothed nailed joints.
■ Spackle tape joints (fill coat).
■ Sand tape joints (as needed).
■ Spackle tape joints (second coat).
■ Sand tape joints (as needed).
■ Spackle tape joints (third time).
■ Sand tape joints (third time).

DR5 INSPECT drywall. Refer to specifications and inspection guidelines. To do this properly, turn out all lights and look at the wall while shining a light on it from an angle. Slight shadows will appear if there are imperfections on the surface. Mark them with a pencil so the drywall sub will know where to fix walls. Don't use a pen as it will leave a mark that will show through paint.

DR6 TOUCH up and repair imperfect drywall areas. There are always a few. Place large drywall scraps in basement if you want to finish the basement later.

DR7 PAY drywall sub. Get him to sign an affidavit.

DR8 PAY drywall sub retainage. You may want to wait until the first coat of paint has been applied so that you can get a good look at the finished product.

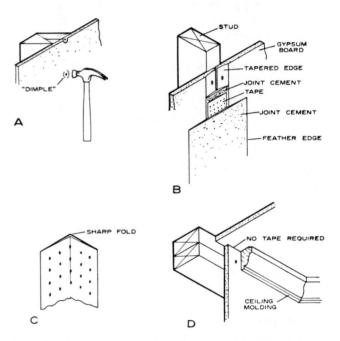

Figure 51. - Finishing gypsum dry wall: A, Nail set with crowned hammer; B, cementing and taping joint; C, taping at inside corners; D alternate finish at ceiling.

DRYWALL: SAMPLE SPECIFICATIONS

■ Bid is to provide all material, labor and equipment to perform complete job per specifications of plans. This includes:
  » Drywall, tape and drywall compound
  » Metal corner bead
  » All nails
  » Sandpaper
  » Ladders and scaffolding

■ Apply 5/8" gypsum board (drywall), double nailed at top. Four nails per stud.

■ All joints to be taped with three separate (3) coats of joint compound each sanded smooth to the touch.

■ All outside corners to be reinforced with metal corner bead.

■ All inside corners to be reinforced with joint tape.

- All necessary electrical outlet, switch and fixture cutouts to be made.

- All necessary HVAC ductwork cutouts to be made.

- All ceilings to be stippled as specified.

- All ceiling sheets to be glued and screwed in place.

- Wall adhesive to be applied to all studs prior to applying drywall.

- Moisture resistant gypsum to be used along all wall areas around shower stalls and bath tubs.

- Use drywall stilts at your own risk.

DRYWALL: INSPECTION

BEFORE TAPING

_____ No more than a 3/8" gap between sheets.

_____ Nails driven in pairs (2 inches apart). Nail heads should be dimpled below the surface of the drywall.

_____ Nails not hit so hard as to break the surface paper. All joints shall be double nailed or glued.

_____ No nail heads exposed as to interfere with drywall.

_____ No sheet warpage, bowing or damage. Sheets are easier to replace before taping.

_____ Rough cuts around door and window openings cut close so that trim will fit properly.

_____ Waterproof drywall, or wonderboard, installed in shower stalls and around bath tubs. No taping is necessary here.

_____ Metal bead installed flush on all outside corners.

DURING FINISHING

_____ Three separate coats of mud are applied to all joints. Some subs only apply two coats to the ceiling if it is to be stippled. The stippling will hide any imperfections. Each successive coat should leave a wider track and a smoother finish.

AFTER FINISHING

_____ Looking down the length of installed drywall, there is no warpage or bumps. If found, circle the area with a pencil (ink may show through when painted). Have the drywall sub-contractor fix all marked areas before final payment.

_____ All joints feathered smooth. No noticeable bumps, either by sight or touch.

_____ All electrical wiring remains exposed including bath and kitchen vent fans, garage door opener switches and doorbell in garage.

_____ Proper ceilings smooth and stippled as specified.

_____ Clean cuts around register openings, switches and outlets so that covers will cover exposed area.

_____ Nap of paper not raised or roughened by excessive or improper sanding.

_____ All touch-up work completed and satisfactory.

Note: Some drywall imperfections will not appear until the first coat of paint has been applied. This is the first time you will see the wall as a single, uniform color. If you can, wait until this point to pay retainage.

DRYWALL: GLOSSARY

BAZOOKA--Term for a automated drywall application tool that tapes, applies mud and feathers in one pass. Gets its name because it looks like a bazooka. Expensive to buy or rent - only used on large scale operations.

DRYWALL --Term used to describe sheets of gypsum.

FEATHERING--Successive coats of drywall compound applied to joints. Each successive pass should widen the compound joint.

FLOAT--To spread drywall compound smooth.

GYPSUM BOARD--Same as DRYWALL

HANG--To hang drywall is to nail it in place.

MUD--Slang for spackle, or drywall compound. Used to seal joints and hide nail head dimples.

SPACKLE --Soft putty-like compound used for drywall patching and touch up that does not shrink as much as regular joint compound. Allows painting immediately after application.

STIPPLE--Rough and textured coatings applied to ceilings. This process makes it easier to finish ceilings since they do not have to be taped and sanded as many times as the walls.

TAPE--Paper used to cover the joints between sheets of gypsum. Tape joints are then sealed with mud (spackle).

23. Trim and Stain

Good trim requires some of the most precision work to be found in a home. If you are not real experienced in this type of work, you may not want to tackle this job. Interior trim subs usually will do the following:

- Set interior/exterior doors and door sills
- Set windows and window sills
- Install base, crown, chair rail, picture and other molding
- Install paneling, raised paneling and wainscoting
- Install stairway trim
- Install fireplace and main entry door mantels
- Install closet shelves and hanger rods
- Install any other special trim

Trim subs charge for work in lots of different ways:

- By the opening (for doors, windows and entryways)
- By the cut. The more cuts involved, the greater the cost.
- By the hour or day.
- By the job.

It is advised to get the bid for the job. Have them give you a bottom line figure - let them do the figuring. Go with the bottom line price and quality you are most comfortable with.

Exterior trim (Facia and cornice work) is covered under a separate section since it is usually done by the siding and cornice sub.

Decorative trim is one of the most visible items in your house and can make the difference between an ordinary interior and one that stands out. The workmanship of your trim sub will be very visible and hard to repair if not done properly. Make sure your trim sub is very quality conscious and watch his work closely. If you want to keep a close eye on his work, volunteer to assist him in doing the trimwork.

The critical part of door installation is in making sure that the door is set level and square so that doorknob hardware will work properly and the door will operate without binding. Special care should also be taken with bifold and sliding track doors. Improper installation is hard to fix and results in binding.

Trimwork has become somewhat easier with the advent of preset doors and window trim kits. To save money, consider using pre hung hollow core interior doors. These doors install easily and look as good or better than custom hung doors.

In choosing your trim material, you must first decide whether you plan to paint or stain the trim. You may be able to use lesser grades of millwork or masonite doors if they will be painted. Finger joint trim consisting of scraps glued together is fine if you plan to paint. Masonite panel doors are becoming very popular and are inexpensive. If you plan to stain, you must use solid wood trim and birch paneled wood doors; but the fine appearance and low maintenance of stained trim may be worth the expense.

When installing base molding, consider whether trim will go around wall registers or merely be interrupted by the register. The first example is more expensive but yields a nicer appearance. If you use a tall base molding, you can build your registers and receptacles into the base molding.

Modern houses are making use of decorative trim extensively such as paneling and wainscoting. Ask your trim sub for suggestions of decorative approaches. It is amazing what can be done with a minimum of materials

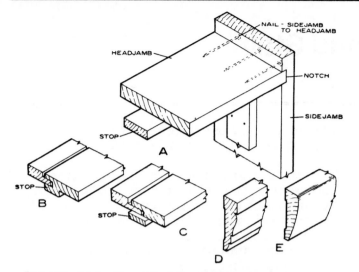

Figure 52. - Interior door parts: A, Door jambs and stops; B, two-piece jamb; C, three piece jamb; D, Colonial casing; E, ranch casing.

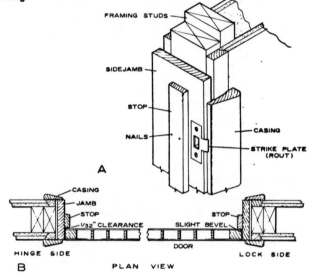

Figure 53. - Door details: A, installation of strike plate; B, location of stops.

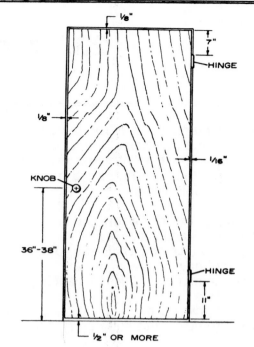

Figure 54. - Door clearances.

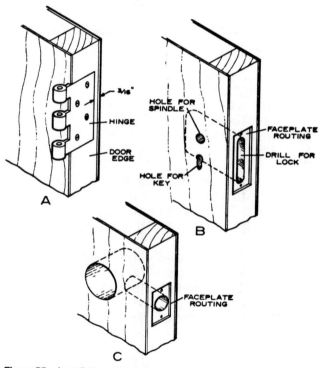

Figure 55. - Installation of door hardware: A, Hinge; B, mortise lock; C, bored lock set.

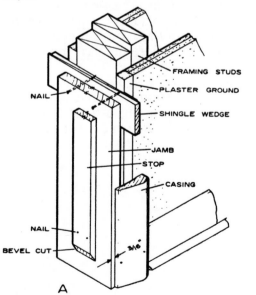

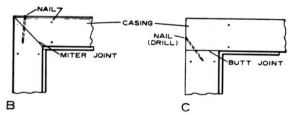

Figure 56. - Doorframe and trim; A, Installation; B, miter joint for casing; C, butt joint for casing.

TRIM: STEPS

TR1 DETERMINE trim requirements.

TR2 SELECT molding, window frames and door frames.

TR3 CONDUCT standard bidding process.

TR4 INSTALL interior doors. If pre-hung doors, install and adjust door stops and doors.

TR5 INSTALL window casing and aprons.

TR6 INSTALL trim around cased openings.

TR7 INSTALL staircase molding. Treads, risers, railings, newels, baluster, goosenecks, etc...

TR8 INSTALL crown molding. This includes special-made inside and outside corners if specified. This is installed first so that the ladder legs will not scratch the base molding.

TR9 INSTALL base and base cap molding.

TR10 INSTALL chair rail molding.

TR11 INSTALL picture molding.

TR12 INSTALL, sand and stain paneling.

TR13 CLEAN all sliding door tracks.

TR14 INSTALL thresholds and weather stripping on exterior doors and windows.

TR15 INSTALL shoe molding after flooring is installed. Painting the shoe molding before installation will reduce painting and touch up work.

TR16 INSTALL door knobs, deadbolts, door stops, and window hardware. Rekey locks if necessary.

TR17 INSPECT trim work. Refer to your specifications and the checklist on the following page.

TR18 CORRECT any imperfect trim and stain work.

TR19 PAY trim sub.

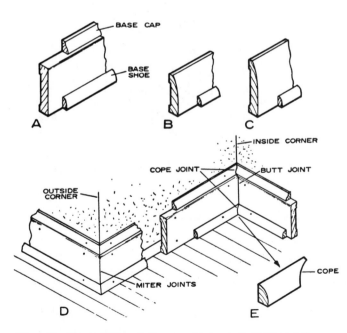

Figure 61. - Base molding; A, Square-edge base; B, narrow ranch base; C, wide ranch base; D, installation; E, cope.

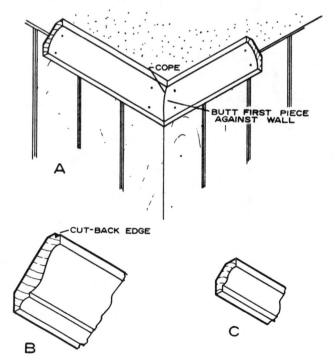

Figure 62. - Ceiling moldings: A, Installation (inside corner); B, crown molding; C, small crown molding.

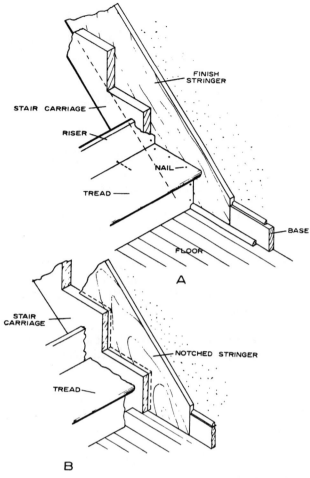

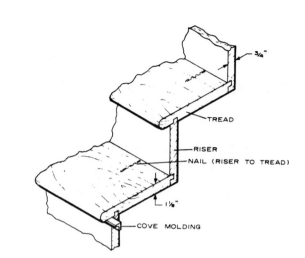

Figure 58. - Main stair detail with combination of treads and risers.

Figure 57. - Enclosed stairway details: A, With full stringer, B, with notched stringer.

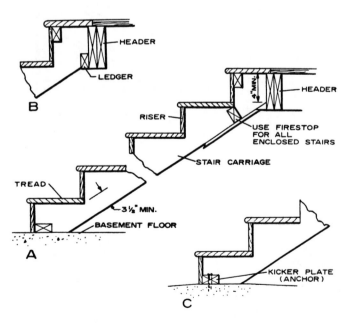

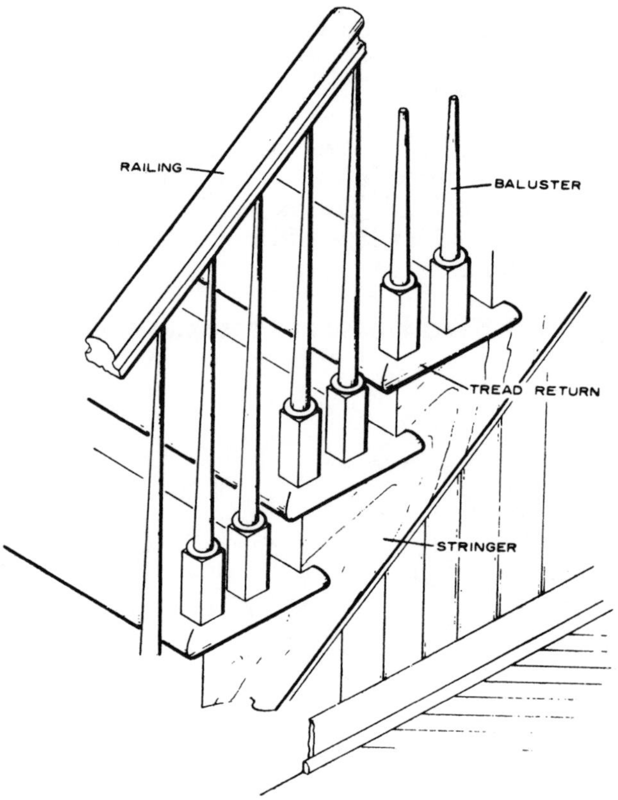

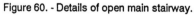

Figure 60. - Details of open main stairway.

Figure 59. - Basement stairs: A, Carriage details; B, ledger for carriage; C, kicker plate.

TRIM: SAMPLE SPECIFICATIONS

- All materials except tools, to be furnished by builder.

- Interior doors will be pre-hung.

- Finish nails to be used exclusively and set below the surface, puttied and sanded over smooth.

BID IS TO INCLUDE THE FOLLOWING:

- Hang all interior doors. All doors to be right and left handed as indicated on attached drawings.

- Install all door and window trim according to attached schedule.

- Install base molding and base cap molding as specified.

- Install apron and crown molding in living room, dining room and foyer area.

- Install two piece chair rail molding in dining room as specified.

- Install 1" shoe molding in all rooms not carpeted.

- Install den bookshelves as indicated on attached drawings.

- Install staircase trim as indicated below:

  - » 12 oak treads
  - » 12 pine risers
  - » solid oak railing with involute
  - » gooseneck
  - » 36 spindles.

- Build fireplace mantel as indicated on attached drawings.

- Install door locks, door knobs, window hardware and dead bolts.

- All trim will be paint grade unless otherwise stated.

- Install all closet trim, shelves and rods.

TRIM: INSPECTION

DOORS

_____ Proper doors installed. Correct style, size, type, etc.

_____ Doors open and close smoothly and quietly. Hinges do not bind or squeak. Doors should swing freely with no noise or friction against adjoining surfaces. Open doors at 30, 45 and 60 degree angles. Doors should remain where positioned. If not, remove center pin, bend it slightly with a hammer and replace.

_____ Door knobs and latches align with latch insets. All deadbolt locks should align properly. Privacy locks installed on proper side of doors. Passage locks on proper doors.

_____ All exterior doors lock and unlock properly. Locks function freely.

_____ All keys available. Have all locks keyed the same for ease of use.

_____ All doors open in proper direction.

_____ All doors plumb against door jambs. With door slightly open, check for alignment and evenness of opening.

_____ All door casing nails set below surface and filled with putty.

_____ Doorlocks and deadbolts work properly without sticking.

_____ No hammer dents in door or door casing.

_____ Thresholds in place and properly adjusted.

_____ Weather stripping in place.

_____ Proper clearance from floor. Consider height of carpet, pad, tile, etc.

_____ Door stops in proper places.

WINDOWS

_____ Window frame secure in place with leveling wedges or chips.

_____ Windows installed with less than 1/2" gap between wall and frame.

_____ Windows open and close smoothly and easily. Windows should glide easily along tracks. Windows should close evenly and completely.

_____ No hammer dents in window or window casing.

_____ All window casing nails set below surface and sealed with putty.

_____ Weatherstripping in place if specified.

_____ Screens installed if specified.

_____ Trim joints are caulked, sanded smooth and undetectable, both by sight and feel.

_____ Trim intersects with walls, ceilings and floors evenly with no gaps or other irregularities.

_____ All trim is void of major material defects.

_____ All finishing nails set below surface and sealed with wood putty.

_____ All den paneling and shelving installed and finished as specified.

_____ Fireplace mantle installed and finished as specified. Will support heavy load.

_____ Fireplace surround installed and finished as specified.

TRIM: GLOSSARY

APRON--Trim used at base of windows. Also used as base to build out crown molding.

BALUSTER--Vertical trim used in quantity to support staircase railings not supported by a wall.

CAP MOLDING--Trim applied to the top of base molding.

BASE MOLDING--Molding applied at the intersection of the wall and floor.

CASING--Molding of various widths used around door and window openings.

CASING NAILS--Used to apply finish trim and millwork. Nail head is small and is set below the surface of the wood to hide it.

CHAIR RAIL--Molding applied to the walls, normally at hip level.

CROWN MOLDING--Molding applied at the intersection of the ceiling and walls. Normally associated with more formal areas.

DEN PANELING--Paneling comprised normally of sections of stain grade materials, joined with a number of vertical and or horizontal strips of molding.

DENTIL MOLDING--A special type of crown molding with an even pattern of "teeth".

FILLER--Putty or other pasty material used to fill nail holes prior to painting or staining.

FINGER JOINT--Trim comprised of many small scrap pieces by a joint resembling two sets of interlocking fingers. This trim is often used to reduce costs where such trim will be painted.

GOOSE NECK--Section of staircase trim with a curve in it.

INVOLUTE--Curved portion of trim used to terminate a piece of staircase railing. Normally used on traditional homes.

MITER JOINT--A diagonal joint, normally 45 degrees, formed at the intersection of two pieces of molding.

MITER BOX--Special guide and saw used to cut trim lumber at precise angles.

MOLDING--Decorative strips of wood or other material applied to wall joints and surfaces as a decorative accent. Molding does not have any structural value.

QUARTER ROUND--A small strip of molding whose cross section is similar to a quarter of a circle. Used with or without base molding. May be applied elsewhere.

TRAY MOLDING--Special type of crown molding where a large portion of the molding is applied to the ceiling as opposed to the wall.

WAINSCOTING--Paneling and trim applied from the floor to a height of about three feet. Used in dining areas to protect against marks from dining chairs. Popular decorative touch in modern homes.

<u>Notes</u>

24. Painting and Wallcovering

If there is any job you can feel safe about doing yourself, it is painting, and wallpapering. It may be worth several thousand dollars to you.

PAINT
Paints can be purchased in standard colors or custom mixed. Custom mixed paints can't be returned, so be careful with color schemes and order quantities. Try to select a minimum of different colors to avoid having to store a lot of different colors. If you decide to paint yourself, purchase the highest quality paints you can find - but make sure to insist on the builder price. One coat paints are a good choice. New drywall should always receive two coats but the one coat paints guarantee the best coverage. Colors on small samples in a store will appear darker when they are applied to a room. If you plan to stipple ceilings, this should be done before any painting has begun. Although roller application is quick and yields good results, consider renting an electric paint sprayer. You may need to thin the paint if you go this route and you will definitely need some form of eye and nose protection. Keep paint off of areas to be stained or it will show up in the final work. When enamel thickens, thin it with paint thinner.

PAINTING TRIM
Trim includes all molding and door/window trim. Although small in terms of absolute square feet, don't be surprised if this effort takes you as long or longer than painting walls. Trim paint (normally enamel-based), is thicker than wall paint. Be careful to use long, smooth strokes to avoid brush marks. For best results use a high quality china bristle brush with tapered bristles.

SAVING TIME
You can save a significant amount of time if you use a power roller. These units keep adequate paint on the roller with a push of a button. Avoid using paint sprayers for interior walls except for the primer coat. Sprayers do not hide minor blemishes and drywall sanding marks like paint rollers and will give disappointing results.

You can save a tremendous amount of time

painting if you coordinate the paint, drywall and trim process between subs. Apply both coats of wall paint and the first coat of trim paint before the doors and trim are installed. This will allow you to roll paint the walls right up to the door openings without worrying about getting paint on the trim. Put one coat on the doors and trim before installation. These two steps will cut interior painting time in half. Make sure to coordinate this with your trim carpenter first. Many trim subs do not like to work with prepainted trim. It creates an irritating dust when sawed. Test masking tape to be sure that it does not pull off paint when removed.

STAIN
Stain grade moldings and handrails should be sanded with a very fine grit sandpaper before staining. Fill holes with a wood filler that accepts stain or color match filler to stain used. Use a good quality wiping stain and don't leave the stain on too long.

WALLCOVERINGS
Wallpaper is an excellent do it yourself project. Most wallpaper outlet stores will be glad to show you the ins and outs of quality wallpaper selection and installation. Most wallpaper comes in double rolls of 32 feet per roll. Single rolls can be purchased at a premium. Wallpaper has a standard width of 24".

When purchasing wallpaper, make sure all rolls have the same run number. The same number tells you that wallpaper of a certain pattern was all made at the same time, and hence the colors will be exactly the same from roll to roll. Repeat pattern is a big factor in estimating the number of rolls that you will need. Repeat pattern is the width (or length) of the pattern before it starts over. On large repeat patterns, you will waste a lot of paper getting patterns to match up.

Wallpaper comes in unpasted and prepasted forms. Many wallpaper hangers still use a paint roller to roll on a wallpaper glue even on prepasted wallpaper. Wallpaper that will be used for baths should be either foil or vinyl due to the humid environment.

PAINTING: STEPS

PT1 SELECT paint schemes. Minimize number of colors. Try not to rely on paint samples since final colors may vary. Try a test mix on your wall if possible.

PT2 PERFORM standard bidding process. If you are or considering doing the painting yourself, this will help you to determine how much you will might save by doing the job yourself. If you are not up to doing the entire job, consider hiring the painter for the exterior work only and doing the interior work yourself.

PT3 PURCHASE all painting materials if you plan to paint yourself. Materials needed will include:

» Gallon cans of ceiling and wall paint.
» Gallon or quart cans of trim paint
» Can opener
» Paint stir sticks
» Paint tray
» Trim guard
» Rolls of 1" masking tape that doesn't adhere to paint
» Several tubes of trim caulk
» Caulk gun
» Single-edged razor blades and handle
» Six-foot step ladder
» Medium and fine grained sand paper
» Dropcloths
» Paint rags
» Paint roller (fine nap) or power roller (self feeding). Consider getting a professional type roller that has a 20" coverage to save time.
» Paint sprayer
» Good quality paint brushes
» Paint thinner

EXTERIOR

PT4 PRIME and caulk all exterior surfaces: around windows, doors, exterior corners, cornice, and spigot openings. Exterior millwork should be primed immediately after installation.

PT5 PAINT exterior siding, trim, shutters and wrought iron railing. Iron railing should be painted with a rust-retarding paint.

PT6 PAINT all cornice work.

PT7 PAINT gutters if needed.

INTERIOR

PT8 PREPARE painting surfaces. This involves performing the following tasks:

» DUST off all drywall with a dry rag.
» APPLY trim caulk to joint between trim and wall. Smooth it with your finger. Wipe off excess. When dry, sand to a smooth joint. This is an important step which will help to yield professional quality results.
» REPAIR any dents in drywall or moldings with spackle or wood filler.
» SAND all repaired areas smooth. NOTE: If ceiling is to be stippled, you don't need to do a thing to it.

PT9 PAINT prime coat; walls and trim. Any good painting job involves two coats of paint. Before you start, make sure you have excellent lighting available in every room to be painted. Even if you intend to wallpaper, paint a primer coat so that it will be easier to remove the wallpaper if you ever want to later. If you have masonite interior doors, prime them also. Ceiling white makes a good neutral primer. Since there is nothing in the house that you can harm, you need not worry about drop cloths. You can either rent a commercial paint sprayer, purchase your own, or use a roller. For a professional job, lightly sand the walls before the final coat after the primer has dried. Sand trim also.

PT10 PAINT or stipple ceilings. A paint roller with an extension arm makes this job much easier. Wear goggles to protect your eyes. If you have stippled ceilings, you will not be painting them.

PT11 PAINT walls. If you want your switch plates and outlet covers to match your wall color, install them before painting. Start with a small brush and paint around the ceiling, all windows and doors. Then cover the large areas with a roller. Do one wall at a time.

PT12 PAINT or stain trim. Start with the highest trim and work down. Allow the walls to dry first. It will be easier if you use a 1 1/2" sash brush for doing windows and a 2" brush for all other trim. Some painters don't bother to cover all of the little trim in multi-paned windows; they just scrape all the dried paint off with a razor blade.

PT13 REMOVE paint from windows with one sided utility razor blades. Do this only when the paint has completely dried or the soft paint shavings will stick to the surface and make a mess.

PT14 INSPECT paint job for spots requiring touch up. Make sure you are looking at the paint job either in good natural light or under a bright light.

PT15 TOUCH-UP paint job.

PT16 CLEAN-UP. Latex paints will come off with soap and warm water. Enamel, which you may have used on your trim, will require mineral spirits or varsol for removal. Dry out the brushes and store them wrapped up in aluminum foil with the bristles in a smooth position. Place Saran Wrap on the paint cans prior to storing them.

PT17 PAY painter (if applicable) and have him sign an affidavit.

PT18 PAY retainage after final inspection.

PAINTING SAMPLE SPECIFICATIONS

INTERIOR PAINTING

■ Bid is to include all material, labor and tools.

■ Primer coat and one finish coat to be applied to all walls including closet interiors.

■ Walls to be touch sanded after primer has dried.

■ Flat latex to be used on drywall, Gloss Enamel to be used on all trim work.

■ Ceiling white to be used on all ceilings.

■ All trim joints to be caulked and sanded before painting.

■ All paint to be applied evenly on all areas to be painted.

■ Window panes to be cleaned by painter.

■ Painting contractor to clean up after job.

■ Excess paint to remain on site when job is completed.

EXTERIOR PAINTING

■ 5-Year warranty exterior grade latex paint to be used.

■ All exterior trim to be primed with an exterior grade primer.

■ All trim area around windows, doors and corners to be caulked with exterior grade caulk before painting. Caulk color to match paint color.

PAINTING: INSPECTION

_____ Proper paint colors used.

_____ All ceilings and walls appear uniform in color with no visible brush strokes.

_____ Trim painted with a smooth appearance. Gloss or semi-gloss enamel used as specified.

_____ All intersections (ceiling/wall, trim/wall, wall/floor) are sharp and clean. Clean straight line between wall and trim colors.

_____ Window panes free of paint inside and out.

_____ Extra touch up paint left at site.

_____ All exterior areas painted smoothly.

_____ No dried paint drops.

PAINTING: GLOSSARY

ALKYD--A durable synthetic resin used in paints.

BASE COAT--First coat put on drywall. Normally a very light color. Should even be put on prior to wallpaper so that the paper can be easily removed.

ENAMEL--Oil base paint use in high soil areas such as trim and doors.

LATEX--Water soluble paint. Normally recommended because of ease of use for interior work.

SAGGING--Slow dripping of excessively heavy coats of paint.

SEMI-GLOSS--A paint or enamel with insufficient nonvolatile vehicle such that the dried coating has a luster but does not look glossy.

TOUCH SANDING--Very light sanding of prime paint coat.

25. Cabinetry and Countertops

Because of the cabinetry, the kitchen invariably turns out to be the most expensive room in the house. Cabinetry and countertops are used in both the kitchen and baths. Money spent on high quality cabinets can really add value to your home. Cabinetry can be prefabricated or custom built for your installation.. Although custom cabinets are nice, the extra expense and time involved is rarely worth it. There is an abundance of high quality prefab cabinets available today.

When shopping for cabinets, pay attention to space saving devices such as revolving corner units, drop down shelves, and slide out shelves. These options can significantly increase the usable space in your cabinets.

KITCHEN CABINETS
Visit a few kitchen cabinet stores or kitchen design stores to become acquainted with the many functional types of cabinets that are available. Look at sample kitchen layouts to get ideas. Determine your material (wood, formica, etc...), hinge and knob styles and finish. Also importantly, determine how much cabinet space you need. The amount of cabinet space is not as important as how well the space is used. Many American craftsman are turning out impressive reproductions of fine, expensive European models at a fraction of the price. The cabinetry should be an integral part of your kitchen design, complementing appliances, lighting and space. Remember, kitchens and baths sell houses.

BATHROOM CABINETS AND VANITIES
Purchase all of your cabinets from the same place if you can. Not only does this make your cabinet selections more uniform, but you can probably get a better price from your cabinet sub. Always look for bathroom cabinets with drawers for storing toiletries.

COUNTER TOPS AND VANITY TOPS
Counter tops are normally made of formica. But other materials have become popular including butcher block, tile, corian and marble (natural or cultured). Formica is by far the most popular and least expensive. Some companies sell bathroom vanities and countertops with the sink built into the countertop as one piece.

Counter tops should have a standard depth of 24" and may have a 3 or 4 inch splashblock against the wall. Countertops are priced and installed by the foot while special charges may be incurred to cut openings for stoves, sinks and fancy edgework. If you get a separate company to do the counter and vanity tops, which is most likely when cultured marble is used, installation is normally included in the price - but get this confirmed. An advantage to cultured marble tops is the lack of a seam between the top and the sink, no sink bowl to purchase, and an easier job for the plumber.

CABINET SUB
This sub will normally produce or supply all cabinetry and install them along with the countertops. This includes making all the cuts for sinks, stoves, etc. Cabinet subs will normally bid the job at one price for material and labor.

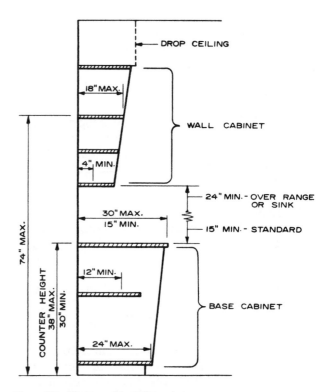

Figure 63. - Kitchen cabinet dimensions.

CABINETRY: STEPS

CB1 CONFIRM kitchen design. Make sure you have enough cabinets to store all pots, pans, utensils and food items.

CB2 SELECT cabinetry/countertop styles and colors.

CB3 COMPLETE diagram of cabinet layout for kitchen and baths.

CB4 PERFORM standard bidding process. There are lots of independent cabinet makers that can copy expensive european cabinet designs.

CB5 PURCHASE or have constructed, all cabinetry and counter top surfaces. If your counter tops come from a separate sub, coordinate the cabinetry and counter top sub. You can't install tops until vanities are in. Unless you know what you are doing, wait for the drywall to be hung before having the cabinet sub measure. This gives them more exact figures to work from and a place to mark on the wall. But do order your cabinets as soon as possible because they can be a source of big delays later. Give your cabinet sub specs for all appliance dimensions.

CB6 STAIN and seal, paint or otherwise finish all cabinet woodwork. Stains will stain the grout and other nearby items, so cabinets must be stained before they are installed.

CB7 INSTALL bathroom vanities. These must be in place before your plumber can connect the sinks, drain pipes and faucets.

CB8 INSTALL kitchen wall cabinets. If you install base cabinets first, they get in your way while installing wall units.

CB9 INSTALL kitchen base cabinets. Also make necessary cutouts for sinks and cooktops. Show the cabinet sub where cutouts are to be made. Don't make any cutouts until the units have arrived.

CB10 INSTALL kitchen and bath countertops, aprons and backsplashes.

CB11 INSTALL and adjust cabinetry hardware. Pulls, hinges, etc...

CB12 INSTALL utility room cabinetry.

CB13 CAULK all wall/cabinetry joints as needed.

CB14 INSPECT all cabinetry and countertops.

CB15 TOUCH up and repair any scratches and other marks on cabinetry and countertops.

CB16 PAY cabinet sub. Have them sign an affidavit.

CABINETRY: SAMPLE SPECIFICATIONS

■ Bid is to provide and complete cabinetry and countertop installation as indicated on attached drawings.

■ All cabinets are to be installed by qualified craftsman.

■ All cabinets are to be installed plumb and square and as indicated on attached drawings.

■ All hardware such as hinges, pulls, bracing and supports to be included in bid.

■ All necessary countertop cut-outs and edging to be included in bid.

■ All exposed corners on countertops to be rounded.

■ Install one double stainless steel sink (model/size).

■ Special interior cabinet hardware includes:

>> Lazy susan
>> Broom rack
>> Pot racks
>> Sliding lid racks

VANITY TOPS SAMPLE SPECIFICATIONS

■ Bid is to manufacture and install the following cultured marble tops in 3/4" solid color material. Sink bowls to be an integral part of top.

■ Vanity A to be 24" deep, 55" wide, finished on left with backsplash; 19" oval sink cut with rounded edge with center of sink 22" from right side. Color is solid white. Spread faucet to be used.

■ Vanity B to be 24" deep, 45" wide, finished on right with backsplash; 19" oval sink cut with rounded edge with center of sink 18" from left side. Color is solid almond. Standard faucet to be used.

CABINETRY: INSPECTION

_____ All cabinet doors open and close properly with no binding or squeaking. Doors open completely and remain in place half opened.

_____ All pulls and other hardware are securely fastened. Check for excessive play in hinges.

_____ All specified shelving is in place and is level.

_____ All exposed cabinetry has an even and smooth finish.

_____ No nicks, scratches, scars or other damage/irregularity on any cabinetry and countertops.

_____ Countertops are level by placing water or marbles on surface. All joints are securely glued with no buckling or delamination.

_____ All drawers line up properly

_____ All mitered and flush joints tight.

_____ Cutouts for sinks and cooktops done properly and units fitted into place.

CABINETRY: GLOSSARY

BACKSPLASH--A small strip (normally three or four inches), placed against the wall and resting on the back of the countertop. This is normally used to protect the wall from water and stains.

BASE CABINET--A cabinet resting on the floor at waist level, supporting a countertop or appliance.

CENTER SET--Standard holes for a standard faucet set.

CORIAN--An artificial material simulating marble manufactured by the Corning ware Company.

LAMINATE--Any thin material, such as plastic or fine wood, glued to the exterior of the cabinet.

SELF-RIMMING--Term used to describe a type of sink that has a heavy rim around the edge that automatically seals the sink against the formica top.

SPREAD SET--Faucet set that requires three holes to be cut wider than normal.

26. Flooring - Tile - Glazing

The most widely used floorings include:

- Carpet
- Hardwood (oak strip, plank and parquet)
- Vinyl floor covering
- Ceramic tile

CARPET is measured, purchased and installed by the square yard and comes in rolls twelve feet wide. Do not try to install carpet yourself unless you have experience; it is not as easy as it appears. Carpet should be installed with a foam pad which adds to the life of the carpet and makes the carpet feel more plush. A stretcher strip will be nailed around the perimeter of the room and the carpet will be stretched flush with the wall. Stretching the carpet prevents it from developing wrinkles. After 2-3 months, the carpet may need to be stretched again. Make sure your carpet company will agree to do this at no cost.

When pricing carpet, always get the installed price with the cost of the foam pad included. A thicker foam pad will make your carpet feel more luxurious but may cost extra. Do not install carpet until all worker traffic has ceased. Make sure that all bumps and trash has been cleaned from the floor before installation. Even the smallest bump will be noticeable through the carpet after installation.

HARDWOOD flooring is measured by the board foot (oak or other plank flooring) or by the square yard (parquet). Oak is laid in tongue and grooved strips nailed to the sub-floor, stained and sealed. Parquet is normally installed as 6" squares with an asphalt adhesive. Parquet squares are available in stained and unstained styles. Avoid laying hardwood floors in humid weather because when the air dries out, gaps will appear between the boards. Buy the flooring a week or two before it is needed and store it in the rooms where it will be laid to let the moisture content stabilize. "Select and better" is the normal grade used in homes. It has a few knots in it. "Character grade" has more knots but is less expensive.

After installation you may want to sand the floor for an extra smooth finish. Pine flooring normally isn't sanded after installation. On all other wood floors, sanding should only be done with the grain or sanding marks will show up.

Use a professional sized drum sander. An "edger" should be used to sand the edges. If you lay hardwood floors on a concrete slab, you need to seal the concrete and use a concrete adhesive. A better method will be to lay down runner strips and nail the floor to the strips. Wood strip floor installers usually work as a pair: one nailing strips in place while the other saws pieces to the proper length and sets them in place. If you can avoid it, do not lay wood floors on slabs.

VINYL FLOOR COVERING is measured by the square yard and comes in a variety of styles and types. Padded vinyl rolls have become very popular because of their ease of installation and maintenance. Installation procedures vary depending on the material and manufacturer, so take care to follow the manufacturers recommendations when installing. Warranties can be voided if the material is not installed properly.

When installing vinyl floors, consider laying an additional subfloor of plywood over the area to be covered. This additional layer provides a smoother surface for the floor and brings the level of the floor up to the same level as any surrounding carpet.

TILE is measured and sole by the square foot although tile units can come in any size. Tile comes either glazed or unglazed and can be installed over a wood sub-floor, although a special adhesive cement and latex grout must be used. Consider using a colored grout which can hide dirt. If you coat the tile grout with a protective coating, you may void the tile warranty.

Tile installation is a tedious and critical procedure. Make sure your tile sub is experienced and quality conscious. Improper installation can result in buckling, cracking and delamination. To insure that tiles adhere properly, avoid installation in cold weather or make sure that the floor is heated for 24 hours after installation.

Colored grouts add a special touch without adding to costs. Don't bother with a colored grout unless the joint is at least 1/8" wide. You may wish to use a grout additive instead of water. This will strengthen the grout and bring out the color. To save time, also consider using a special protective coating applied to tile prior

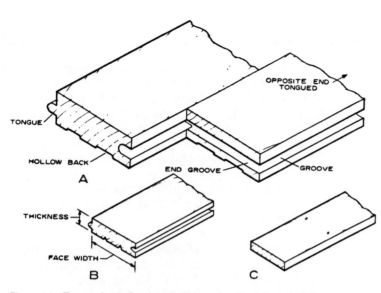

Figure 64. - Types of strip flooring: A, Side and end-notched - 25/32–inch; B, thin flooring strips–matched; C, thin flooring strips–square-edged.

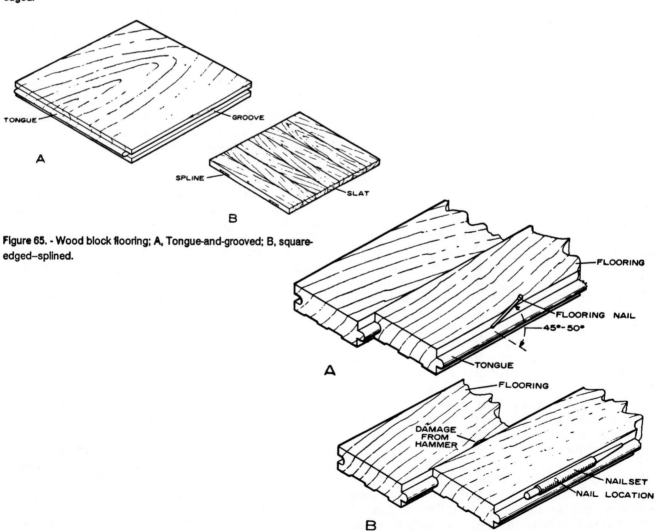

Figure 65. - Wood block flooring; A, Tongue-and-grooved; B, square-edged–splined.

Figure 66. - Nailing of flooring; A, Nail angle; B, setting of nail.

to grouting. This will allow you to wipe the grout film off of the tile easily when the job is done.

Remember that you have a 1/2" fudge factor around floor perimeters if you use shoe molding. If you tile yourself, ask your supplier to loan or rent you a tile cutter with a new blade or buy a new blade - this will make a world of difference.

Tiling shower and bath stalls is tricker than floors. You may want to do the floors, but hire a sub for the rest. Use spacers when tiling the floor to keep the tiles straight.

GLAZING, while not part of flooring, normally occurs during the flooring phase. Glazing refers to any mirrors and plate glass windows that must be custom cut from sheets of plate glass. No special specifications are needed.

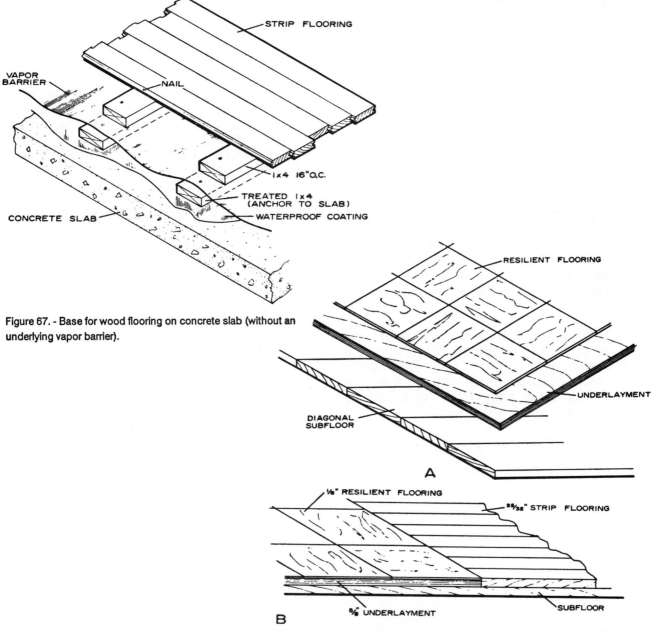

Figure 67. - Base for wood flooring on concrete slab (without an underlying vapor barrier).

Figure 68. - Base for wood flooring and concrete slab (without an underlying vapor barrier).

GLAZING: STEPS

GL1 DETERMINE mirror and special glazing requirements. This means all of the following items:

- Mirrors
- Closet door mirrors
- Shower doors
- Fixed picture windows
- Medicine cabinets

GL2 SELECT mirrors and other items. Here you must decide such trivia as beveled edge or non-beveled edge mirror glass, chrome or gold frame for shower doors, etc... Glazing subs will have catalogs for you to look through.

GL3 CONDUCT standard bidding process for glazing sub.

GL4 INSTALL fixed pane picture windows. Check that window has no scratches and is adequately sealed.

GL5 INSTALL shower doors and medicine cabinets. If you can, turn on shower head and point toward the shower door to check for leaks. Door should glide freely along track.

GL6 INSTALL mirrors and closet door mirrors. Mirrors should be installed with a non-abrasive mastic. Check for hairline scratches in the silver coating; this cannot be fixed aside from replacing the mirror.

GL7 INSPECT all glazing installation.

GL8 CORRECT all glazing problems.

GL9 PAY glazier and have them sign an affidavit.

FLOORING AND TILE: STEPS

FL1 SELECT carpet brands, styles, colors and coverage. Also select your padding. Keep your styles and colors to a minimum. It helps your buying power, minimizes scrap and provides for easier resale.

FL2 SELECT hardwood floor type, brand, style, color and coverage.

FL3 SELECT vinyl floor covering brand, style, color and coverage.

FL4 SELECT tile brands, styles, colors and coverage. This includes tile for bath walls, bath floors, shower stalls, patios and kitchen counter tops. Once your sub-floor is in, borrow about a square yard of your favorite tile from your supplier. Lay it down where it will be used and see how it looks.

FL5 CONDUCT standard bidding process For best prices on carpet, go to a carpet specialist or carpet mill. When possible, have the supplier bid for material and labor. If your supplier does not install, ask for installer references.

FL6 ORDER hardwood flooring. Hardwood flooring should remain in the room in which it will be installed for several weeks in order for it to expand and contract based on local humidity conditions. You may want to order a few extra bundles of shorter, cheaper lengths to do closets and other small areas.

TILE

FL7 ORDER tile and grout. If you do the job yourself, borrow or rent a tile cutter from the tile supplier. Look for good bargains on "seconds"; tile with slight imperfections, to use as cut pieces.

FL8 PREPARE area to be tiled. Sweep floor clean and nail down all squeaky floor areas with ridged nails.

FL9 INSTALL tile base in shower stalls. This normally involves applying concrete over a wire mesh. In shower stalls, this is done on top of the shower pan. If you have tile in the kitchen do not tile under the cabinets or the island. Prior to tiling around a whirlpool tub, make sure the tub is secure and has been fully wired, grounded and plumbed.

FL10 APPLY tile adhesive. Use a trowel with grooves suitable for your tile and grout. Use large curved, sweeping motions.

FL11 INSTALL tile and marble thresholds. Don't forget to use tile spacers to assure even tile spacing. Make sure tubs are secure in place before tiling.

FL12 APPLY grout over tile. Apply silicone sealant between the tub and tile. This will help absorb the vibration from the motor.

FL13 INSPECT tile. Refer to specifications and inspection criteria.

FL14 CORRECT any problems that need fixing.

FL15 SEAL grout. This should only be done after grout has been in place for about three weeks. Use a penetrating sealer to protect the grout from stains.

FL16 PAY tile sub. Get him to sign an affidavit.

FL17 PAY tile sub retainage after floor has proven to be stable and well glued.

HARDWOOD

FL18 PREPARE subfloor for hardwood flooring. If you have a WOOD SUB-FLOOR, you should put down at least two layers of heavy duty building paper. If you are laying hardwood on a slab, you have two choices:

■ Seal the slab with a liquid sealer and install flooring with an adhesive

■ Prepare a subfloor:

» Sweep slab
» Apply 1" x 2" treated wood strips called bottom sleepers with adhesive and 1 1/2" concrete nails, 24" apart, perpendicular to oak strips
» Lay 6 mil poly vapor barrier over strips
» Lay second layer of 1" x 2" wood strips over vapor barrier.
» Nail wood flooring strips to the firring strips.

FL19 INSTALL hardwood flooring. Ask installers to leave the scraps. Store them away for repair work or other use.

FL20 SAND hardwood flooring. Although most floors are pre-sanded, you may still need to have them sanded for the smoothest possible finish. Sand only with the grain. Cross-grain sanding will not look bad until the stain is applied. Once floors are sanded, there should be no traffic until after the sealer is dry.

FL21 INSPECT hardwood flooring. Refer to specifications and inspection guidelines.

FL22 CORRECT any problems with hardwood flooring. Common problems are split strips, hammer dents in strip edges, uneven spacing, gaps, squeaks, non-staggered joints and last strip not parallel with wall.

FL23 STAIN hardwood flooring. Floors must be swept well before performing this step. Close windows if it is dusty outside.

FL24 SEAL hardwood flooring. Polyurethane is the most popular sealant. You normally have an option of either gloss or satin finish. Place construction paper over dry floors to protect them from construction traffic.

FL25 INSPECT hardwood flooring again. This time, primarily for finish work.

FL26 PAY hardwood flooring sub. Have them sign an affidavit.

FL27 PAY hardwood flooring sub retainage.

VINYL FLOOR COVERING

FL28 PREPARE sub-floor for vinyl floor covering. Sweep floor clean and nail down any last squeaks with ridged nails. Plane down any high spots.

FL29 INSTALL vinyl floor covering and thresholds. Ask installers to leave behind any large scraps that can be used for repair work.

FL30 INSPECT vinyl floor covering. Refer to specifications and inspection guidelines.

FL31 CORRECT any problems with vinyl floor covering.

FL32 PAY vinyl floor covering sub. Have him sign and affidavit.

FL33 PAY vinyl floor covering sub retainage.

CARPET

FL34 PREPARE sub-floor for carpeting. Sweep floor clean and nail or plane down down any high spots. This is your last chance to fix squeaky floors easily and install hidden speaker wires.

FL35 INSTALL carpet stretcher strips.

FL36 INSTALL carpet padding. You should use an upgraded pad if you used a tongue and groove sub-floor. If you done use a carpet stretcher on this step you are doing the job wrong.

FL37 INSTALL carpet. This should be done with a carpet stretcher to get a professional job. Ask installer to leave larger scraps for future repair work.

FL38 INSPECT carpet. Refer to specifications and inspection guidelines.

FL39 CORRECT any problems with carpeting installation.

FL40 PAY carpet sub. Have them sign an affidavit.

FL41 PAY carpet retainage.

FLOORING AND TILE: SPECIFICATIONS

CARPET

- Bid is to furnish and install wall-to-wall carpet in the following rooms:
  - » Living room
  - » Dining room
  - » Master Bedroom and closets
  - » Second Bedroom and closets
  - » Third Bedroom and closets
  - » Second Floor Stairway
  - » Second Floor Hallway
  - » 8" accent border to be installed in master bathroom. Color to be of builder's choice.

- Full size upgraded pad is to be provided and installed under all carpet.

- Stretcher strips are to be installed around carpet perimeters.

- All seams to be invisible.

- All major carpet and padding scraps to remain on site.

- All necessary thresholds to be installed.

- All work to be performed in a professional manner.

OAK FLOORING

- Bid is to provide and install 'select and better' grade flooring in the following areas:
  - » Foyer and first floor hallway
  - » Den

- Oak strips to be single width, 3 1/4" strips.

- Two perpendicular layers of heavy gauge paper to be installed between sub-floor and oak flooring.

- All oak strips to be nailed in place with flooring nails at a 50 degree angle. Nails to be installed only with a nailing machine to prevent hammer head damage to floor.

- Shorter lengths to be used in closets.

- All oak strips to be staggered randomly so that no two adjacent joint ends are within 6" of each other.

- Oak flooring and oak staircase treads to be sanded to a smooth surface.

- Oak flooring and oak staircase treads to be swept, stained, sealed and coated with satin finish polyurethane.

- Stain to be a medium brown.

- Ridged or spiral flooring nails to be used exclusively.

- Only first and last strip to be surface-nailed, concealable by shoe mold.

PARQUET FLOORING

- Bid is to provide and install parquet flooring in the following areas:
  - » Sunroom
  - » Sunroom closet

- Parquet flooring to be stained and waxed.

- Install marble thresholds in the following areas:
  - » Between hallway and living room
  - » Between hallway and dining room
  - » Between hallway and den
  - » Between kitchen and dining room

VINYL FLOOR COVERING

- Bid is to provide and install vinyl floor covering in the following areas:
  - » Kitchen and kitchen closet
  - » Utility room

- All seams to be invisible.

- All major scraps to remain on site.

- All necessary thresholds to be installed.

TILE

- Bid is to provide and install specified tile flooring in the following areas:
  - » Master bath shower stall walls. One course of curved base tile around perimeter. Tile to go all the way to ceiling.
  - » Master bath shower floor.
  - » Surround and one step for whirlpool bath.
  - » Second bath shower stall walls all the way to ceiling.
  - » Second bath floor (per diagram).
  - » Half bath floor. Per attached diagram. One course of base tile around perimeter.
  - » Kitchen floor (per diagram). Tan grout sealed. One course of base tile around perimeter.
  - » Kitchen wall backsplash. See drawings attached.
  - » Kitchen island top.
  - » Hobby room and closet.

- Grout colors to be as specified by builder.

- All necessary soap dishes, tissue holders, towel racks and tooth brush holders to be supplied and installed in white unless otherwise specified by builder in writing.

- Matching formed base and cap tiles to be used as needed to finish edging and corners.

- 1/4" per foot slope toward drain in shower stalls.

- All necessary tub and shower drains installed.

- All grout to be sealed three weeks after grout is installed.

FLOORING AND TILE: INSPECTION

CARPET

_____ Proper carpet is installed.

_____ No visible seams.

_____ Carpet is stretched tightly and secured by carpet strips.

_____ Colors match as intended. Watch for color variations

_____ Carpet installed on stairs fits tight on each riser.

OAK FLOORING AND PARQUET

_____ Proper oak strips and/or parquet is installed. (Make, color, etc...)

_____ No flooring shifts, creaks or squeaks under pressure.

_____ Even staining and smooth surface.

_____ No scratches, cracks or uneven surfaces. Scratches normally indicate sanding across the grain or using too coarse of sandpaper for final pass.

_____ Smaller lengths used in closets to reduce scrap and waste.

_____ All necessary thresholds are installed and are parallel to floor joints.

_____ Only edge boards have visible nails. No other boards have visible nails.

_____ Edge boards are parallel to walls.

_____ No irregular or excessive spacing between boards.

_____ No floor registers covered up by flooring.

VINYL FLOOR COVERING

_____ Proper vinyl floor covering is installed. (Make, color, pattern, etc...)

_____ Vinyl floor covering has smooth adhesion to floor. No bubbles.

_____ No movement of flooring when placed under pressure.

_____ No visible or excessive seams.

_____ No scratches or irregularities.

_____ All necessary thresholds have been installed and are parallel to floor joints.

_____ All necessary thresholds are of proper height for effective door operation. No more than 1/8" gap under interior doors.

_____ Pattern is parallel or perpendicular to walls as specified.

TILE

_____ Proper tiles and grouts are installed. (Make, style, dimensions, color, pattern, etc...)

_____ Tile joints are smooth, evenly spaced, parallel and perpendicular to each other and walls where applicable.

_____ Tiles are secure and do not move when placed under pressure.

_____ No scratches, cracks, chips or other irregularities.

_____ Grout is sealed as specified.

_____ Grout lines parallel and perpendicular with walls.

_____ Tile pattern matches artistically.

_____ Tiles are laid, symmetrically from the center of the room out. Cut pieces on opposite sides of the room are of the same size.

_____ All tile cuts are smooth and even around perimeter and floor register openings.

_____ Marble thresholds are installed level with no cracks or chips.

_____ Marble thresholds parallel with door when closed.

_____ Marble thresholds are at proper height.

_____ Matching formed tiles used at base of tile walls as specified.

_____ Cap tiles used at top of tile walls.

_____ Proper tile edging for tubs, sinks and shower stalls.

_____ No cracked or damaged tiles.

_____ Tile floor is level.

<u>Notes</u>

27. Landscaping

LANDSCAPING

When you consider how much resale value and curb appeal a good landscaping job can add to a home, it is surprising that so many builders scrimp on this important effort. Many builders just throw up a few bushes and then seed or sod. Hopefully you will be willing to spend just a bit more time and effort to do the job right. If you don't want to spend all the money at once, you can devise a two or three-year implementation plan, beginning with the critical items such as ground coverage.

Important items to keep in mind are:

- Plan for curb appeal. Your lawn should be a showplace.

- Plan for low maintenance. If you don't want to trim hedges and rake leaves forever, be smart about the trees and bushes you install or keep.

- Work with the sun; not against it. Don't force grass to grow in the shade of trees or the home. Maximize sunlight into the home by careful planning of trees.

- Plan for proper drainage. This is CRITICAL! Water must drain away from dwelling and not collect in any low spots.

- Consider doing much of this work yourself.

- Consider brief assistance from a landscape architect.

Every lot has one best use and many poor ones. A good landscape architect can help you do more than select grass. Landscape architects can be helpful in:

- Determining best position of the home on the lot.

- Determining attractive driveway patterns.

- Determine which trees should remain and additional trees and/or bushes to plant for privacy and lowest maintenance.

- Determine soil condition and necessary additives.

- Determine proper drainage pattern and ground coverings.

You can normally pay an independent landscape architect by the hour, but if you do a lot of work with him, work up a fixed price. Meet him at your lot with the lot survey. Walk the lot with him, getting his initial impressions and ideas. From there, you may request a site plan. The site plan will normally be a large drawing showing exact home position, driveway layout, existing trees and bushes, sculptured islands, and other topographical features.

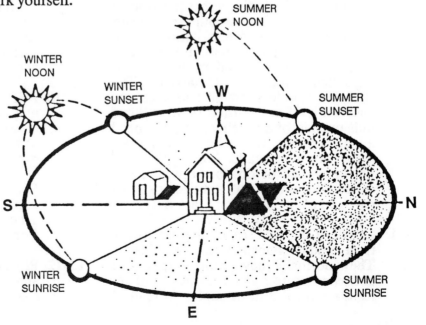

□ excellent location
▦ good location
▓ poor location

LANDSCAPING: STEPS

BEFORE CONSTRUCTION

LD1 EVALUATE your lot in terms of trees to keep, drainage, privacy, slopes, high and low spots and exposure to morning and afternoon sun. A landscape architect can help you greatly with this step. He can also help you determine if you will have too little or too much soil after excavation and what to do about it. Excavated soil may be used to fill in low spots or otherwise assist in developing a drainage pattern or landscaping feature.

LD2 DETERMINE best possible position for home on lot. Considerations should include:

- Good Southern exposure for parts of the house a lot of glass.

- Best view for most important rooms when at all possible. Don't give your best view to a garage or windowless wall.

- Best angle for curb appeal. Consider angling your home if on a corner lot.

LD3 DEVELOP site plan. This is a birds-eye view of what you want the finished site to look like, including exact position of home, drive way and all other paved surfaces such as walks and patios, major trees, islands, grass areas, flower beds, bushes and even an underground watering system if you intend to have one. The site plan should show all baselines and easements. If any portion of your home falls outside of a baseline, you will need a land variance. When determining grass areas, consider shade. Most grasses do not grow well or do not grow at all in shade. No amount of fertilizer or lawn care can compensate for a lack of sun.

LD4 CONTACT a landscape architect to critique and/or improve on your design. They normally have more experience and can surprise you with new and interesting ideas. For example, instead of having a driveway go straight from the street to the garage, consider putting a lazy curve in the drive and a slight mound on the street side. This helps hide the expanse of concrete which is not normally a focal point. Landscape architects can recommend specific types of bushes to suit special privacy needs.

LD5 FINALIZE site plan. Make several copies in blue-print form. You will need at least one for yourself, one for a landscaper (if used) and one for the county.

LD6 SUBMIT site plan to building officials and bank along with your blue prints.

LD7 PAY landscape architect (if used one).

AFTER CONSTRUCTION

LD8 CONDUCT soil tests. Soil samples should be made for every several thousand feet of yard. Soil to be tested should be taken from six inches down. Soil testing will help to determine PH (level of acidity) and the lack of essential nutrients. Most counties have an extension service which will test the samples for you at little or no charge. Good general purpose soil conditioners are peat moss, manure and lime.

LD9 TILL first four to six inches of soil, adding soil conditioners as needed. Cut any unwanted trees down and grind stumps before sod is laid down. Rake surface smooth first one way and then another.

LD10 APPLY soil treatments as needed. You may also with to add a slow release, non-burning fertilizer (no stronger than 6-6-6) to help your young grass grow. Apply gypsum pellets to break down clay.

LD11 INSTALL underground sprinkling system as needed. This normally consists of a network of 1/2" PVC pipe purchased from a garden supply store.

LD12 PLANT flower bulbs. This is the easiest and best time to plant daffodils, crocus, hyacinths, gladiolas, etc...

LD13 APPLY seed or sod, depending upon your choice. Seed is much cheaper, but sod means an instant thriving lawn and elimination of topsoil runoff. Seed is hard to start on a hill or in other adverse conditions. If you do chose seed, use a broadcast spreader or a special seed dispensing device. Cover the seed with hay or straw to protect seeds from washing away with rain. If you sod, this will be a tougher job, but doing it yourself could really save some money. Sod always has at least three prices: direct from a sod farm, delivered by a sod broker or delivered and installed by a sod broker. If you can, buy direct from a sod farm. Visit them if you have time to see before you buy. Have them drop the sod in shady areas all around the grounds to be covered to reduce carrying. Keep sod slightly moist until installed. Have a good crew to help you lay sod because it will only live on pallets for a few days. Do not install sod before your water is hooked up.

LD14 SOAK lawn with water if sod is applied. Sprinkle lightly for long periods of time if a seeded lawn is used. Roll lawn with a heavy metal cylinder.

LD15 INSTALL bushes and trees. If you plan to install trees, please keep them at least four feet from the driveway area. Trees are notorious for breaking up a driveway when the roots begin to spread.

LD16 PREPARE landscaped islands. Place pine bark, pine straw, mulch, or gravel in designated areas. Over time, pine straw will kill some grasses particularly bermuda. Keep pine straw in islands only.

LD17 INSTALL mailbox. This is only if you have the kind that sits on a post. If you have a masonry or large frame mailbox, these will be installed during the framing or masonry stages of construction.

LD18 INSPECT landscaping job. Refer to specifications and checklist.

LD19 CORRECT any problems noted.

LD20 PAY landscaping sub for job. Have them sign an affidavit.

LD21 PAY landscaping sub retainage after grass is fully established without bare areas.

LANDSCAPING SPECIFICATIONS

■ Bid is to landscape lot per site plan as described below:

■ Loosen soil in yard to a depth of six inches with no dirt clumps larger than six inches.

■ Finely rake all area to be seeded or sodded.

■ Install the following soil treatments:

Sphagnum peat moss: 21 cubic feet per 1000 square feet
Sand: 500 lbs. per 1000 square feet
Lime: 200 lbs. per 1000 square feet
6-6-6: as prescribed by manufacturer for new lawns

■ All grass to be lightly fertilized with 6-6-6 and watered heavily immediately to prevent drying.

■ Sod to be bermuda 419, laid within 48 hours of delivery.

■ Sod pieces to butt each with no gaps with joints staggered.

■ Islands to be naturally shaped by trimming sod.

■ Sod to be rolled after watering.

■ Prepare three pine straw islands as described in attached site plan.

■ Install trees and bushes according to attached planting schedule and site plan.

■ Cover all seeded areas with straw to prevent runoff

■ Install six splash blocks for downspouts.

LANDSCAPING: INSPECTION

_____ All areas sodded/seeded as specified.

_____ All trees and bushes installed as specified.

_____ Sod and/or seed is alive and growing.

_____ Sod is smooth and even with no gaps.

_____ Straw covers seed evenly with no bare spots.

_____ All specified bushes and trees are alive and in upright position.

_____ Splash blocks installed.

_____ Grass cut if specified.

_____ No weeds in yard.

_____ No exposed dirt within 16" of siding or brick veneer.

_____ Pine bark or pine straw islands installed as specified.

28. Sources of Additional Information

The following sources are generally recognized as members of the building industry which have impact on standards and guidelines.

Aluminum Association (AA)
818 Connecticut Avenue, N.W.
Washington D.C. 20006

American Concrete Institute (ACI)
P.O. Box 19150
Detroit, Michigan 48219

American Hardboard Association
205 West Tough Avenue
Park Ridge, Illinois 60611

American Insurance Association (AIA)
85 John Street
New York, New York 10020

American Institute of Steel Construction, Inc.
1221 Avenue of the Americas, Suite 1580
New York, New York 10020

American Iron and Steel Institute (AISI)
1000 16th Street, N.W.
Washington, D.C. 20036

American Institute of Timber Construction (AITC)
333 West Hampden Avenue
Englewood, Colorado 80110

American National Standards Institute (ANSI)
1430 Broadway
New York, New York 10018

American Plywood Association (APA)
1119 A Street
Tacoma, Washington 98401

Asphalt Roofing Industry Bureau (ARIB)
750 Third Avenue
New York, New York 10017

American Society of Heating, Refrigeration and Air Conditioning Engineers (ASHRAE)
United Engineering Center
345 East 47th Street
New York, New York 10017

American Soc. for Testing Materials (ASTM)
1916 Race Street
Philadelphia, Pennsylvania 19103

American Wood Preservers Association (AWPA)
1625 I Street, N.W.
Washington, D.C. 20006

American Wood Preservers Institute (AWPI)
1651 Old Meadow Road
Arlington, Virginia 22101

American Wood Preservers Bureau (AWPB)
P.O. Box 6085
Arlington, Virginia 22206

American Welding Society, Inc. (AWS)
2501 N.W. 7th Street
Miami, Florida 33125

Brick Institute of America (BIA)
1750 Old Meadow Road
McLean, Virginia 22101

National Clay Pipe Institute (NCPI)
1130 17th Street, N.W.
Washington, D.C. 20036

Concrete Reinforcing Steel Institute (CRSI)
180 North LaSalle Street
Chicago, Illinois 60601

Commercial Standards
(U.S. Department of Commerce)
Superintendent of Documents
Government Printing Office
Washington, D.C. 20402

Forest Products Laboratory (FPL)
United States Department of Agriculture
Madison, Wisconsin 53705

Federal Specifications (FS)
Superintendent of Documents
Government Printing Office
Washington, D.C. 20234

Gypsum Association (GA)
1603 Orrington Avenue
Evanston, Illinois 60201

The Hydronics Institute (HI)
35 Russo Place
Berkely Heights, New Jersey 07922

Hardwood Plywood Manufacturer's Association (HPMA)
P.O. Box 6246
Arlington, Virginia 22206

International Masonry Industry All-Weather Council (IMIAWC)
208 South LaSalle Street
Chicago, Illinois 60604

Metal Building Manufacturers Association (MBMA)
2130 Keith Building
Cleveland, Ohio 44115

Manufacturing Chemists' Association, Inc.
1825 Connecticut Avenue, N.W.
Washington, D.C. 20009

Mobile Homes Manufacturers Association (MHMA)
14650 Lee Road
Chantilly, Virginia 22021

Metal Lath Association (MLA)
221 N. LaSalle Street, Suite 1507
Chicago, Illinois 60601

National Association of Building Manufacturers (NABM)
1619 Massachusetts Avenue, N.W.
Washington, D.C. 20036

National Association of Home Builders (NAHB)
15th & M Streets, N.W.
Washington, D.C. 20005

National Automatic Sprinkler and Fire Control Association, Inc. (NASFCA)
45 Kensico Drive
Mt. Kisco, New York 10549

National Bureau of Standards (NBS)
(U.S. Department of Commerce)
Superintendent of Documents
Government Printing Office
Washington, D.C. 20402

National Concrete Masonry Association (NCMA)
2302 Horse Pen Road, P.O. Box 781
Herndon, Virginia 22070

National Environment Systems Contractors Association (NESCA)
1501 Wilson Boulevard
Arlington, Virginia 22209

National Fire Protection Association (NFPA)
470 Atlantic Avenue, N.W.
Washington, D.C. 20036

National Forest Products Association
1619 Massachusetts Avenue, N.W.
Washington, D.C. 20036

National Lime Association (NLA)
4000 Brandywine Street, N.W.
Washington, D.C. 20016

National Mineral Wool Association, Inc. (NMWAI)
382 Springfield Avenue - Suite 312 Basset Bldg.
Summit, New Jersey 07901

National Particleboard Association (NPA)
2306 Perkins Place
Silver Springs, Maryland

National Pest Control Association (NPCA)
250 West Jersey Street
Elizabeth, New Jersey 07202

Portland Cement Association (PCA)
5420 Old Orchard Road
Skokie, Illinois 60076

Prestressed Concrete Institute (PCI)
20 N. Wacker Drive
Chicago, Illinois 60606

Product Fabrication Service (PFS)
1618 West Beltline Highway
Madison, Wisconsin 53713

Perlite Institute (PI)
45 West 45th Street
New York, New York 10036

Post-Tensioning Institute (PTI)
Suite 3500, 301 West Osborn
Phoenix, Arizona 85013

Pittsburgh Testing Laboratory (PTL)
1330 Locust Street
Pittsburgh, Pennsylvania 85013

Red Cedar Shingle and Handsplit Shake Bureau (RDSHSB)
5510 White Building
Seattle, Washington 98101

Southern Gas Association (SGA)
4230 LBJ Freeway, Suite 414
Dallas, Texas 75234

Steel Joist Institute (SJI)
703 Parham Road
Richmond, Virginia 23229

Sheet Metal & Air Conditioning Contractor's National Association (SMACCNA)
8224 Old Court House Road
Vienna, Virginia 22180

Southern Forest Products Association (SFPA)
P.O. Box 52468
New Orleans, Louisiana 70152

Society of Plastics Industry, Inc. (SPI)
355 Lexington Avenue
New York, New York 10017

Southern Pine Inspection Bureau (SPIB)
P.O. Box 846
Pensacola, Florida 32502

Steel Window Institute (SWI)
2130 Keith Building
Cleveland, Ohio 44115

Southwest Research Institute (SRI)
8500 Culebra Road
San Antonio, Texas 78228

Tile Council of America (TCA)
P.O. Box 326
Princeton, New Jersey 08540

Truss Plate Institute (TPI)
1800 Pickwick Avenue
Glenville, Illinois 60025

Underwriters' Laboratories, Inc. (ULI)
333 Pfingsten Road
Northbrook, Illinois 60062

Western Wood Products Association (WWPA)
1500 Yeon Building
Portland, Oregon 97204

<u>Notes</u>

Appendices

Masterplan

PRE-CONSTRUCTION

PC1   Begin construction project scrapbook
PC2   DETERMINE size of home
PC3   EVALUATE alternate house plans
PC4   SELECT house plan
PC5   MAKE design changes
PC6   PERFORM material and labor takeoff
PC7   CUT costs as required
PC8   UPDATE material and labor takeoff
PC9   ARRANGE temporary housing
PC10 FORM separate corporation
PC11 APPLY for business license
PC12 ORDER business cards
PC13 OBTAIN free and clear title to land
PC14 APPLY for construction loan
PC15 ESTABLISH project checking account
PC16 CLOSE construction loan
PC17 OPEN builder accounts
PC18 OBTAIN purchase order forms
PC19 OBTAIN building permit
PC20 OBTAIN builder's risk policy
PC21 OBTAIN workman's compensation policy
PC22 ACQUIRE minimum tools
PC23 VISIT construction sites
PC24 DETERMINE need for on-site storage
PC25 NOTIFY all subs and labor
PC26 OBTAIN compliance bond
PC27 DETERMINE minimum occupancy requirements
PC28 CONDUCT spot survey of lot
PC29 MARK all lot boundaries and corners
PC30 POST construction signs
PC31 POST building permit on site
PC32 ARRANGE for portable toilet
PC33 ARRANGE for site telephone

EXCAVATION

EX1   CONDUCT excavation bidding process
EX2   LOCATE underground utilities
EX3   DETERMINE exact dwelling location
EX4   MARK area to be cleared
EX5   MARK where curb is to be cut
EX6   TAKE last picture of site area
EX7   CLEAR site area
EX8   EXCAVATE trash pit
EX9   INSPECT cleared site
EX10 EXCAVATE foundation/basement area
EX11 INSPECT site and excavated area
EX12 APPLY crusher run to driveway area

NOTES

EX13 SET UP silt fence
EX14 PAY excavator (initial)
EX15 PLACE supports on foundation wall
EX16 INSPECT backfill area
EX17 BACKFILL foundation
EX18 CUT driveway area
EX19 COVER septic tank and septic tank line
EX20 COVER trash pit
EX21 PERFORM final grade
EX22 CONDUCT final inspection
EX23 PAY excavator (final)

PEST CONTROL
PS1 CONDUCT standard bidding process.
PS2 SCHEDULE pest control sub for pre-treatment
PS3 APPLY pesticide to foundation
PS4 APPLY pesticide to perimeter
PS5 OBTAIN signed pest control warranty
PS6 PAY pest control sub

CONCRETE
CN1 CONDUCT bidding process (concrete supplier)
CN2 CONDUCT bidding process (formwork/finishing)
CN3 CONDUCT bidding process (block masonry)
CN4 INSTALL batter boards
CN5 INSPECT batter boards
CN6 DIG footings
CN7 SET footing forms
CN8 INSPECT footing forms
CN9 SCHEDULE and complete footing inspection
CN10 CALL concrete supplier
CN11 POUR footings
CN12 INSTALL key forms
CN13 FINISH footings as needed
CN14 REMOVE footing forms
CN15 INSPECT footings
CN16 PAY concrete supplier for footings
CN17 SET poured wall concrete forms
CN18 SCHEDULE concrete supplier
CN19 INSPECT poured wall concrete forms
CN20 POUR foundation walls
CN21 FINISH poured foundation walls
CN22 REMOVE poured wall concrete forms
CN23 INSPECT poured walls
CN24 BREAK OFF tie ends from walls
CN25 PAY concrete supplier for walls

NOTES

CN26 ORDER blocks/schedule masons
CN27 LAY concrete blocks
CN28 FINISH concrete blocks
CN29 INSPECT concrete block work
CN30 PAY concrete block masons
CN31 CLEAN up after block masons
CN32 SET slab forms
CN33 PACK slab sub-soil
CN34 INSTALL stub plumbing
CN35 SCHEDULE HVAC sub for slab work
CN36 INSPECT plumbing and gas lines
CN37 POUR and spread crushed stone
CN38 INSTALL vapor barrier
CN39 LAY reinforcing wire as required
CN40 INSTALL rigid insulation
CN41 CALL concrete supplier for slab
CN42 INSPECT slab forms
CN43 POUR slab
CN44 FINISH slab
CN45 INSPECT slab
CN46 PAY concrete supplier for slab
CN47 COMPACT drive and walkway soil
CN48 SET drive and walkway forms
CN49 CALL concrete supplier
CN50 INSPECT drive and walkway forms
CN51 POUR drive and walkway
CN52 FINISH drive and walkway areas
CN53 INSPECT drive and walkway areas
CN54 ROPE OFF drive and walkway areas
CN55 PAY concrete supplier
CN56 PAY concrete finishers
CN57 PAY concrete finisher's retainage

WATERPROOFING
WP1 CONDUCT standard bidding process
WP2 SEAL footing and wall intersection
WP3 APPLY first coat of portland cement
WP4 APPLY second coat of portland cement
WP5 APPLY waterproofing compound
WP6 APPLY 6 mil poly vapor barrier
WP7 INSPECT waterproofing (tar/poly)
WP8 INSTALL gravel bed for drain tile
WP9 INSTALL drain tile
WP10 INSPECT drain tile
WP11 APPLY top layer of gravel on drain tile
WP12 PAY waterproofing sub
WP13 PAY waterproofing sub retainage

FRAMING
FR1 CONDUCT standard bidding process (material)
FR2 CONDUCT standard bidding process (labor)

NOTES

FR3   DISCUSS aspects of framing with crew
FR4   ORDER special materials
FR5   ORDER first load of framing lumber
FR6   INSTALL sill felt
FR7   ATTACH sill plate with lag bolts
FR8   INSTALL support columns
FR9   SUPERVISE framing process
FR10   FRAME first floor joists and subfloor
FR11   FRAME basement stairs
FR12   POSITION all first floor large items
FR13   FRAME exterior walls (first floor)
FR14   PLUMB and line first floor
FR15   FRAME second floor joists and subfloor
FR16   POSITION all large second floor items
FR17   FRAME exterior walls (second floor)
FR18   PLUMB and line second floor
FR19   INSTALL ceiling joists
FR20   FRAME roof
FR21   INSTALL roof decking
FR22   PAY framing first installment
FR23   INSTALL lapped tar paper
FR24   INSTALL pre fab fireplaces
FR25   FRAME dormers and skylights
FR26   FRAME chimney chases
FR27   FRAME tray ceilings, skylights and bays
FR28   INSTALL sheathing on exterior walls
FR29   INSPECT sheathing
FR30   REMOVE temporary bracing
FR31   INSTALL exterior windows and doors
FR32   APPLY deadwood
FR33   INSTALL roof ventilators
FR34   FRAME decks
FR35   INSPECT framing
FR36   SCHEDULE loan draw inspection
FR37   CORRECT any framing problems
FR38   PAY framing labor (second payment)
FR39   PAY framing retainage

ROOFING

RF1   SELECT roofing style, material and color
RF2   CONDUCT standard bidding process.
RF3   ORDER shingles and roofing felt
RF4   INSTALL metal drip edge
RF5   INSTALL roofing felt
RF6   INSTALL drip edge on rake
RF7   INSTALL roofing shingles
RF8   INSPECT roofing
RF9   PAY roofing subcontractor
RF10   PAY roofing subcontractor retainage

NOTES

GUTTERS

GU1 SELECT gutter type and color
GU2 PERFORM standard bidding process
GU3 INSTALL underground drain pipe
GU4 INSTALL gutters and downspouts
GU5 INSTALL splashblocks
GU6 INSPECT gutters
GU7 PAY gutter subcontractor
GU8 PAY gutter sub retainage

PLUMBING

PL1   DETERMINE type/quantity of fixtures
PL2   CONDUCT standard bidding process
PL3   ORDER special plumbing fixtures
PL4   APPLY for water connection and sewer tap
PL5   INSTALL stub plumbing in foundation
PL6   MOVE large plumbing fixtures into place
PL7   MARK location of all fixtures
PL8   INSTALL water meter and spigot
PL9   INSTALL sewer tap
PL10 INSTALL rough-in plumbing
PL11 SCHEDULE plumbing inspection
PL12 INSTALL septic tank and line
PL13 CONDUCT rough-in plumbing inspection
PL14 CORRECT rough-in plumbing problems
PL15 PAY plumbing sub for rough-in
PL16 INSTALL finish plumbing
PL17 TAP into water supply
PL18 CONDUCT finish plumbing inspection
PL19 CORRECT finish plumbing problems
PL20 PAY plumbing sub for finish work
PL21 PAY plumbing retainage

HVAC

HV1   CONDUCT energy audit
HV2   SHOP for heating and cooling
HV3   CONDUCT standard bidding process
HV4   FINALIZE HVAC design
HV5   INSTALL rough-n heating and air
HV6   INSPECT heating and air rough-in
HV7   CORRECT rough-in problems
HV8   PAY HVAC sub for rough-in
HV9   INSTALL finish heating and air
HV10 CONDUCT final HVAC inspection
HV11 CORRECT all HVAC problems
HV12 CALL gas company to install lines
HV13 DRAW gas line on plat diagram
HV14 PAY HVAC sub for finish work
HV15 PAY HVAC retainage

NOTES

ELECTRICAL

EL1   DETERMINE electrical requirements
EL2   SELECT electrical fixtures/appliances
EL3   DETERMINE phone installation re-
        quirements
EL4   CONDUCT standard bidding process
EL5   SCHEDULE phone wiring
EL6   APPLY for temporary electrical hookup
EL7   INSTALL temporary electric pole
EL8   PERFORM rough-in electrical
EL9   INSTALL phone wiring
EL10 SCHEDULE rough-in electrical inspec-
        tion
EL11 INSPECT rough-in electrical
EL12 CORRECT rough-in electrical problems
EL13 PAY electrical sub for rough-in work
EL14 PERFORM finish electrical work
EL15 INSPECT finish electrical
EL16 CORRECT finish electrical problems
EL17 CALL electrical utility to connect ser-
        vice
EL18 PAY electrical sub for finish work
EL19 PAY electrical sub retainage

INSULATION

IN1   DETERMINE insulation requirements
IN2   CONDUCT standard bidding process.
IN3   INSTALL exterior wall insulation
IN4   INSTALL soundproofing
IN5   INSTALL floor insulation
IN6   INSTALL attic insulation
IN7   INSPECT insulation
IN8   CORRECT insulation work as needed
IN9   PAY insulation sub
IN10 PAY insulation sub retainage

DRYWALL

DR1   CONDUCT standard bidding process.
DR2   ORDER and receive drywall
DR3   HANG drywall on all walls
DR4   FINISH drywall
DR5   INSPECT drywall
DR6   TOUCH UP and repair drywall
DR7   PAY drywall sub
DR8   PAY drywall sub retainage

MASONRY

MA1   DETERMINE brick pattern and color
MA2   DETERMINE stucco pattern and color
MA3   PERFORM standard bidding process
        (masonry)

NOTES

MA4 PERFORM standard bidding process (stucco)
MA5 APPLY flashing
MA6 LAY bricks
MA7 FINISH bricks and mortar
MA8 INSPECT brickwork
MA9 CORRECT brickwork deficiencies
MA10 PAY masons
MA11 CLEAN UP excess bricks
MA12 INSTALL base for decorative stucco
MA13 PREPARE stucco test area
MA14 APPLY stucco
MA15 APPLY first coat of stucco
MA16 APPLY second coat of stucco
MA17 INSPECT stucco work
MA18 CORRECT stucco deficiencies
MA19 PAY stucco sub
MA20 PAY mason sub retainage
MA21 PAY stucco sub retainage

SIDING/CORNICE
SC1   SELECT siding material and color
SC2   CONDUCT standard bidding process
SC3   INSTALL windows
SC4   INSTALL flashing
SC5   INSTALL siding
SC6   INSTALL siding trim around corners
SC7   CAULK siding joints
SC8   INSPECT siding
SC9   PAY sub for siding work
SC10 INSTALL cornice
SC11 INSPECT cornice work
SC12 CORRECT any cornice problems
SC13 PAY cornice sub
SC14 CORRECT siding problems
SC15 PAY siding retainage
SC16 PAY retainage

TRIM
TR1   DETERMINE trim requirements
TR2   SELECT molding and millwork
TR3   CONDUCT standard bidding process
TR4   INSTALL interior doors
TR5   INSTALL window casing and aprons
TR6   INSTALL trim around cased openings
TR7   INSTALL staircase openings
TR8   INSTALL crown molding
TR9   INSTALL base and base cap molding
TR10 INSTALL chair rail molding
TR11 INSTALL picture molding
TR12 INSTALL, sand and stain paneling
TR13 CLEAN all sliding door tracks

NOTES

TR14   INSTALL thresholds and weather strip-
       ping
TR15   INSTALL shoe molding
TR16   INSTALL door and window hardware
TR17   INSPECT trimwork
TR18   CORRECT trimwork
TR19   PAY trim sub

PAINTING
PT1   SELECT paint schemes.
PT2   PERFORM standard bidding process.
PT3   PURCHASE all painting materials
PT4   PRIME and caulk exterior surfaces
PT5   PAINT exterior siding and trim
PT6   PAINT cornice work
PT7   PAINT gutters
PT8   PREPARE painting surfaces
PT9   PRIME walls and trim
PT10   PAINT or stipple ceilings
PT11   PAINT walls
PT12   PAINT or stain trim
PT13   REMOVE paint from windows
PT14   INSPECT paint job
PT15   TOUCH UP paint job
PT16   CLEAN UP
PT17   PAY painter
PT18   PAY retainage

CABINETRY
CB1   CONFIRM kitchen design
CB2   SELECT cabinetry/countertop
CB3   COMPLETE cabinet layout diagram
CB4   PERFORM standard bidding process
CB5   PURCHASE cabinetry and countertops
CB6   STAIN or paint cabinetry
CB7   INSTALL bathroom vanities
CB8   INSTALL kitchen wall cabinets
CB9   INSTALL kitchen base cabinets
CB10   INSTALL kitchen and bath countertops
CB11   INSTALL cabinet hardware
CB12   INSTALL utility room cabinetry
CB13   CAULK cabinet joints as needed
CB14   INSPECT cabinetry and countertops
CB15   TOUCH UP and repair as needed
CB16   PAY cabinet sub

FLOORING AND TILE
FL1   SELECT carpet color, style and cover-
       age
FL2   SELECT hardwood floor type
FL3   SELECT vinyl floor covering

NOTES

FL4   SELECT tile type and coverage
FL5   CONDUCT standard bidding process
FL6   ORDER hardwood flooring
FL7   ORDER tile and grout
FL8   PREPARE area to be tiled
FL9   INSTALL tile base in shower stalls
FL10 APPLY tile adhesive
FL11 INSTALL tile and marble thresholds
FL12 APPLY grout over tile
FL13 INSPECT tile
FL14 CORRECT tile problems
FL15 SEAL grout
FL16 PAY tile sub
FL17 PAY tile sub retainage
FL18 PREPARE sub floor for hardwood flooring
FL19 INSTALL hardwood flooring
FL20 SAND hardwood flooring
FL21 INSPECT hardwood flooring
FL22 CORRECT hardwood flooring problems
FL23 STAIN hardwood flooring
FL24 SEAL hardwood flooring
FL25 INSPECT hardwood flooring (final)
FL26 PAY hardwood flooring sub
FL27 PAY hardwood flooring sub retainage
FL28 PREPARE sub floor for vinyl floor covering
FL29 INSTALL vinyl floor covering
FL30 INSPECT vinyl floor covering
FL31 CORRECT vinyl floor covering problems
FL32 PAY vinyl floor covering sub
FL33 PAY vinyl floor covering sub retainage
FL34 PREPARE subfloor for carpeting
FL35 INSTALL carpet stretcher strips
FL36 INSTALL carpet padding
FL37 INSTALL carpet
FL38 INSPECT carpet
FL39 CORRECT carpet problems
FL40 PAY carpet sub
FL41 PAY carpet retainage

NOTES

GLAZING

GL1   DETERMINE glazing requirements
GL2   SELECT mirrors and other items
GL3   CONDUCT standard bidding process
GL4   INSTALL picture windows
GL5   INSTALL shower doors
GL6   INSTALL mirrors
GL7   INSPECT glazing installation
GL8   CORRECT all glazing problems
GL8   PAY glazing contractor

LANDSCAPING

LD1   EVALUATE lot
LD2   DETERMINE best position
LD3   DEVELOP site plan
LD4   CONTACT landscape architect
LD5   FINALIZE site plan
LD6   SUBMIT site plan
LD7   PAY landscape architect
LD8   CONDUCT soil test
LD9   PREPARE topsoil
LD10 APPLY soil treatments
LD11 INSTALL underground sprinkling system
LD12 PLANT flower bulbs
LD13 APPLY seed or sod
LD14 SOAK lawn with water
LD15 INSTALL bushes and trees
LD16 PREPARE landscaped islands
LD17 INSTALL mailbox
LD18 INSPECT landscaping job
LD19 CORRECT any landscaping problems
LD20 PAY landscaping sub
LD21 PAY landscaping sub retainage

NOTES

Project Schedule

HOW TO USE THE PROJECT SCHEDULE

The Project Schedule that follows on the next two pages gives you a visual representation of your entire project over a span of eighteen months.

The time span of your project may vary considerably from this length, so adjust your timing accordingly. The length of each project bar includes some slack time and starts at the earliest possible date. The shaded boxes represent the critical path of the project. In other words, each preceding task must be completed before the next CRITICAL TASK can begin. Step numbers correspond directly to the steps in the Project Management section and in the Master Plan. Each dash between the vertical lines separating weeks equals one day.

MAJOR STEPS		0	1	2	3	4
PRECONSTRUCTION	PC	1-33				
EXCAVATION	EX	1	2-14			
PEST CONTROL	PS	1-2		3		
CONCRETE	CN	1-3		4-9	10-46	
WATERPROOFING	WP	1				2-13
FRAMING	FR	1-4				5
ROOFING	RF	1-2				
PLUMBING	PL	1-4		5		
HVAC	HV	1-4				
ELECTRICAL	EL	1-6		7		
MASONRY	MA	1-4				
SIDING & CORNICE	SC	1-2				
INSULATION	IN	1-2				
DRYWALL	DR	1				
TRIM	TR	1-3				
PAINTING	PT	1-2				
CABINETRY	CB	1-4				
FLOORING & TILE	FL	1-5				
GLAZING	GL	1-3				
GUTTERS	GU	1-2				
LANDSCAPING	LD	1-7				

WEEK OF PROJECT													
5	6	7	8	9	10	11	12	13	14	15	16	17	18

15-17   18-23

4-6

47-57

6-33   34-39

3   4-9   10

6   7-15   16-21

5-8   9-15

8-13   14-19

5-21

3-4   5-9   10-16

3-5   6-10

2-8

4-12   13-19

3-4   5-6   7   8-18

5   6-16

6-27   28-41

4   5-9

3-8

8-21

Notes

SUPPLIERS/SUBCONTRACTORS REFERENCE SHEET

Date _____ How contacted? _____

Name _____ Phone _____

Address_____ Call when? _____

Type of product _____

REFERENCES	PHONE	COMMENTS
_____	_____	_____
_____	_____	_____
_____	_____	_____

PRICES/COMMENTS _____
_____
_____
_____

Date _____ How contacted? _____

Name _____ Phone _____

Address_____ Call when? _____

Type of product _____

REFERENCES	PHONE	COMMENTS
_____	_____	_____
_____	_____	_____
_____	_____	_____

PRICES/COMMENTS _____
_____
_____
_____

Date _____ How contacted? _____

Name _____ Phone _____

Address_____ Call when? _____

Type of product _____

REFERENCES	PHONE	COMMENTS
_____	_____	_____
_____	_____	_____
_____	_____	_____

PRICES/COMMENTS _____
_____
_____
_____

SUPPLIERS/SUBCONTRACTORS REFERENCE SHEET

Date _____ How contacted? _____

Name _____ Phone _____

Address_____ Call when? _____

Type of product _____

REFERENCES   PHONE   COMMENTS

_____   _____   _____

_____   _____   _____

_____   _____   _____

PRICES/COMMENTS _____

_____

_____

_____

_____

Date _____ How contacted? _____

Name _____ Phone _____

Address_____ Call when? _____

Type of product _____

REFERENCES   PHONE   COMMENTS

_____   _____   _____

_____   _____   _____

_____   _____   _____

PRICES/COMMENTS _____

_____

_____

_____

_____

Date _____ How contacted? _____

Name _____ Phone _____

Address_____ Call when? _____

Type of product _____

REFERENCES   PHONE   COMMENTS

_____   _____   _____

_____   _____   _____

_____   _____   _____

PRICES/COMMENTS _____

_____

_____

_____

GENERAL TELEPHONE LIST

CITY/COUNTY

NAME	CONTACT/TITLE	PHONE	HOURS

FINANCIAL INSTITUTION

NAME	CONTACT/TITLE	PHONE	HOURS

UTILITIES

NAME	CONTACT/TITLE	PHONE	HOURS
GAS:			
ELECTRIC:			
WATER:			
TELEPHONE:			

SUBCONTRACTOR TELEPHONE LIST

SUB/LABOR NAME	CONTACT	PHONE	BID AMT.
EXCAVATION:_____	_____	_____	_____
CONCRETE:_____	_____	_____	_____
PESTCONTROL:_____	_____	_____	_____
MASONRY:_____	_____	_____	_____
FRAMING:_____	_____	_____	_____
INSULATION:_____	_____	_____	_____
ELECTRICAL:_____	_____	_____	_____
PLUMBING:_____	_____	_____	_____
HVAC:_____	_____	_____	_____
ROOFING:_____	_____	_____	_____
LANDSCAPING:_____	_____	_____	_____
DRYWALL:_____	_____	_____	_____
TRIM:_____	_____	_____	_____
PAINTING:_____	_____	_____	_____
GUTTERS:_____	_____	_____	_____
CABINETRY:_____	_____	_____	_____
_____	_____	_____	_____
_____	_____	_____	_____
_____	_____	_____	_____
_____	_____	_____	_____
_____	_____	_____	_____
_____	_____	_____	_____
_____	_____	_____	_____
_____	_____	_____	_____
_____	_____	_____	_____
_____	_____	_____	_____
_____	_____	_____	_____
_____	_____	_____	_____
_____	_____	_____	_____
_____	_____	_____	_____
	TOTAL:		_____

PLAN ANALYSIS CHECKLIST

PROPERTY:_____

PLAN NUMBER: _____

CIRCULATION:

Is there access between the following without going through another room?

	YES	NO
Living Room to Bathroom	YES____	NO____
Living Room to Bathroom	YES____	NO____
Family Room to Bathroom	YES____	NO____
Each Bedroom to Bathroom	YES____	NO____
Kitchen to Dining Room	YES____	NO____
Living Room to Dining Room	YES____	NO____
Kitchen to Outside Door	YES____	NO____
Can get from auto to Kitchen without getting wet if raining	YES____	NO____

ROOM SIZE AND SHAPE:

	YES	NO
Are all rooms adequate size and reasonable shape based on planned use?	YES____	NO____
Can Living Room wall accommodate sofa and end tables (minimum 10 ft.)?	YES____	NO____
Can wall of Den take sofa and end tables (minimum 10 ft.)?	YES____	NO____
Can wall of master Bedrooms(s) take double bed and end tables?	YES____	NO____

STORAGE AND CLOSETS:

	YES	NO
Guest closet near front entry	YES____	NO____
Linen closet near baths	YES____	NO____
Broom closet near kitchen	YES____	NO____
Tool and lawnmower storage	YES____	NO____
Space for washer and dryer	YES____	NO____

EXTERIOR:

	YES	NO
Is the house style "normal" architecture?	YES____	NO____
Is the house style compatible with surrounding architecture?	YES____	NO____

SITE ANALYSIS:

	YES	NO
Does the house conform to the neighborhood?	YES____	NO____
Is the house suited to the lot?	YES____	NO____
Is the topography good?	YES____	NO____
Will water drain away from the house?	YES____	NO____
Will grade of driveway be reasonable?	YES____	NO____
Is the lot wooded?	YES____	NO____
Is the subject lot as good as others in the neighborhood?	YES____	NO____

ITEM ESTIMATE WORK SHEET

VENDOR NAME	ITEM NO.	DESCRIPTION	MEASURING QTY. UNIT	CONVERSION FACTOR	ORDERING QTY. UNIT	PRICE EACH	COST	TAX	TOTAL COST	COST TYPE

EXPENSE CATEGORY	MATERIAL COST	LABOR COST	SUBCONT. COST	TOTAL COST

COST ESTIMATE SUMMARY

	EST. COST	ACTUAL COST	OVER/UNDER
LICENSES AND FEES			
PREPARATION OF LOT			
FOOTINGS			
FOUNDATION			
ADDITIONAL SLABS			
MATERIALS (PACKAGE)			
FRAMING LABOR			
TRIM LABOR			
ROOFING			
HVAC			
ELECTRIC			
INTERIOR FINISH (SHEETROCK)			
PAINTING			
FLOOR COVERING			
WALKS, DRIVES, STEPS			
LANDSCAPING			
CABINETS AND VANITIES			
APPLIANCES AND LIGHTS			
INSULATION			
CLEANING			
DEVELOPED LOT COST			
CONSTRUC. CLOSING AND INTEREST			
SUPERVISORY FEE			
REAL ESTATE COMMISSION			
POINTS AND CLOSING			
CONTINGENCY (MISC. EXPENSES)			
SEPTIC			
FIREPLACE			
GUTTERS			
TOTAL			

COST ESTIMATE CHECKLIST

CODE NO. DESCRIPTION	QTY.	MATERIAL UNIT PRICE	MATERIAL TOTAL MAT'L	LABOR UNIT PRICE	LABOR TOTAL LABOR	SUBCONTR. UNIT PRICE	SUBCONTR. TOTAL SUB	TOTAL

TOTALS:

COST ESTIMATE CHECKLIST

CODE NO. DESCRIPTION	QTY.	MATERIAL UNIT PRICE	TOTAL MAT'L	LABOR UNIT PRICE	TOTAL LABOR	SUBCONTR. UNIT PRICE	TOTAL SUB	TOTAL

TOTALS:

PURCHASE ORDER

INV. NO. _____

VENDOR CODE _____

JOB NO. _____

CAT. NO. _____

TO:

DELIVER TO:

TERMS: PAGE ____ OF ____   DATE:

TRANSACTION	INVOICE NO.	ACCOUNT NO.	SALESMAN	CUST P.O. NO.	REFERENCE	

B/O	QTY.	ITEM CODE	DESCRIPTION	UNITS	PRICE/UNIT	TOTAL	CAT.
					NET SALE		
					SALES TAX		
					TOTAL		

PURCHASE ORDER CONTROL LOG

P.O. NO.	DATE ORDERED	DESCRIPTION	VENDOR NAME	DATE DEL	DISCOUNT RATE	DISCOUNT DUE	CK.#	AMT.	PAYMENTS CK.#	PAYMENTS AMT	CK.#	AMT	TOTAL PAID

DESCRIPTION OF MATERIALS

JOB _____ DATE: _____

JOB DESCRIPTION: _____

1. FOUNDATION VENEER: Stucco ____   Rock ____   Brick ____   Other ____

   Description: _____ ____

2. EXTERIOR SIDING: Cedar ____   Pine ____   Brick ____   Masonite ____
   Stucco ____   Other ____

   Description: _____

3. ROOF: Cedar Shakes ____   Shingles ____

   Description: _____

4. FIREPLACE: Prefab ____   Masonry ____   Other ____

   Veneer: Brick ____   Rock ____   Other ____
   Mantel: Yes ____   No ____

   Description: _____

5. HEAT/AC: F/A gas ____   Heat pump ____   Electric ____   Other ____
   Single ____   Dual ____

   Description: _____
   Capacity: _____

6. PLUMBING:

   Master bathroom: Fib. tub ____   Tile ____   Color _____
   Hall bathroom: Fib. tub ____   Tile ____   Color _____
   Guest bathroom: Fib. tub ____   Tile ____   Color _____
   Other: Fib. tub ____   Tile ____   Color _____

   Brand fixtures: _____

   Description: _____

7. FLOORING:

Carpet: Where/Brand/color _____
                      Where/Brand/color _____
Vinyl: Where/Brand/color _____
                      Where/Brand/color _____
Hardwood: Where/Brand/color _____
Ceramic tile: Where/Brand/color _____
Other: Where/Brand/color _____

8. DECK/PATIO:

Deck ____   Patio ____   Steps ____   Size ____

Material:_____

9. INSULATION:

Exterior Walls: Type ____   R Factor ____
Ceiling: Type ____   R Factor ____
Wall sheathing: Type ____   R Factor ____
Other: Type ____   R Factor ____

10. WINDOWS:

Type: _____
Special: _____

11. DRIVEWAY:

Brick ____   Concrete ____

Description: _____

12. WALKS:

Brick ____   Concrete ____

Description: _____

13. TRIM:

Type _____   Size _____   Where _____
Type _____   Size _____   Where _____
Type _____   Size _____   Where _____
Type _____   Size _____   Where _____

14. PANELLING:

Type _____   Color _____   Where _____

15. PAINT:

Interior:
Color _____ Brand _____ Where _____
Color _____ Brand _____ Where _____
Color _____ Brand _____ Where _____
Color _____ Brand _____ Where _____

Exterior:
Color _____ Brand _____ Where _____
Color _____ Brand _____ Where _____
Color _____ Brand _____ Where _____

16. WALLPAPER:
Style _____ Brand _____ Where _____
Style _____ Brand _____ Where _____
Style _____ Brand _____ Where _____
Style _____ Brand _____ Where _____

17. COUNTER TOPS:
Type _____ Color _____ Style No _____ Where _____
Type _____ Color _____ Style No _____ Where _____
Type _____ Color _____ Style No _____ Where _____
Type _____ Color _____ Style No _____ Where _____

18. CABINETS:

Kitchen: Color _____ Brand _____ Style _____
Master Bath: Color _____ Brand _____ Style _____
Hall Bath: Color _____ Brand _____ Style _____
Other: Color _____ Brand _____ Style _____

19. STORMS/SCREENS: Yes _____   No _____   Where _____

20. SHUTTERS: Yes _____   No _____   Where _____

21. DOORS:
Type _____ Brand _____ Where _____
Type _____ Brand _____ Where _____
Type _____ Brand _____ Where _____

_____
BUYER

_____
BUYER

_____
CONTRACTOR

_____
DATE

LIGHTING AND APPLIANCE ORDER

JOB _____

DATE_____

LIGHTING BUDGET_____

DESCRIPTION	QTY.	STYLE NO.	UNIT PRICE	TOTAL PRICE
Front Entrance				
Rear Entrance				
Dining Room				
Living Room				
Den				
Family Room				
Kitchen				
Kitchen sink				
Breakfast Area				
Dinette				
Utility Room				
Basement				
Master Bath				
Hall Bath 1				
Guest Bath				
Hall				
Hall				
Stairway				
Master Bedroom				
Bedroom 2				
Bedroom 3				
Bedroom 4				
Closets				
OUTDOOR:				
Front Porch				
Rear Porch				
Porch/Patio				
Carport/Garage				
Post/Lantern				
Floodlights				
Chimes				
Pushbuttons				
APPLIANCES				
Oven				
Hood				
Dishwasher				
Disposal				
Refrigerator				
SIGNATURE			TOTAL:	

SUBCONTRACTOR'S AGREEMENT

JOB NO. _____   DESCRIPTION _____

ORDER DATE _____   PLAN NUMBER _____

NAME _____   HAVE WORKMANS COMP. YES ___ NO ___

_____   WORKMANS COMP. NO. _____

_____   EXPIRATION DATE _____

PHONE _____   TAX ID NO. _____

PAYMENT TO BE MADE AS FOLLOWS:

_____   _____

_____   _____

_____   _____

DEDUCT _____ % FOR RETAINAGE.

All material is guaranteed to be as specified. All work to be completed in a timely manner acccccording to standard practices. Any alteration or deviation from specifications below involving extra costs will be executed only upon written order (refer to Change Authorization Form).

SPECIFICATIONS: (Attach extra sheets if necessary.)

AGREEMENT _____   BID AMOUNT _____

_____   BID GOOD UNTIL _____
Subcontractor   Date

DATE	CHECK NO.	AMOUNT PAID	MOUNT DUE
_____	_____	_____	_____
_____	_____	_____	_____
_____	_____	_____	_____
_____	_____	_____	_____

SUBCONTRACTOR'S AFFIDAVIT

STATE OF: _____

COUNTY OF: _____

Personally appeared before the undersigned attesting officer, _____
_____, who being duly sworn, on oath says that he was the contractor
in charge of improving the property owned by
_____,
located at _____.

Contractor says that all work, labor, services and materials used in such improvements were
furnished and performed at the contractor's instance; that said contractor has been paid or partially
paid the full contract price for such improvements; that all work done or material furnished in
making the improvements have been paid for at the agreed price, or reasonable value, and there are
...o unpaid bills for labor and services performed or materials furnished; and that no person has any
claim or lien by reason of said improvements.

_____
(Seal)

Sworn to and subscribed before me,
this _____ day of _____, 19_____.

_____
Notary Public

BUILDING CHECKLIST

JOB _____

DATE STARTED _____   ESTIMATED COMPLETION DATE _____

(**Bold faced** numbered items indicate critical tasks that must be done before continuing.)

		DATE DONE:	NOTES:

_____   **A. PRE CLOSING**
_____ 1. Take off   _____ _____
_____ 2. Spec sheet   _____ _____
_____ 3. Contract   _____ _____
_____ 4. Submit loan   _____ _____
_____ 5. Survey   _____ _____
_____ 6. Insurance   _____ _____
_____ 7. Closing   _____ _____

_____   **B. POST CLOSING**
_____ 1. Business License   _____ _____
_____ 2. Compliance Bond   _____ _____
_____ 3. Sewer Tap   _____ _____
_____ 4. Water Tap   _____ _____
_____ 5. Perk test (if necessary) _____ _____
_____ 6. Miscellaneous Permits _____ _____
_____ 7. Building Permit   _____ _____
_____ 8. Temporary Services:
          A. Electrical   _____ _____
_____   B. Water   _____ _____
_____   C. Phone   _____ _____

_____   **C. CONSTRUCTION**
_____ 1. Rough layout   _____ _____
_____ 2. Rough grading   _____ _____
_____ 3. Water
          A. Meter   _____ _____
_____   B. Well   _____ _____
_____ 4. **Set temporary pole**   _____ _____
_____ 5. Crawl or basement
          A. Batter boards   _____ _____
_____   **B. Footings dug**   _____ _____
_____   **C. Footings poured**   _____ _____
_____   D. Lay Block/Pour   _____ _____
_____ 6. Pretreat   _____ _____

DATE DONE: **NOTES:**

_____   7.   Slab
_____   A. Form boards   _____ _____
_____   **B. Plumbing**   _____ _____
_____   C. Miscellaneous Pipes   _____ _____
_____   D. Gravel   _____ _____
_____   E. Poly   _____ _____
_____   **F. Re-Wire**   _____ _____
_____   G. Pour concrete   _____ _____
_____   8.   Spot survey   _____ _____
_____   9.   Waterproof   _____ _____
_____   10.   Fiberglass tubs   _____ _____
_____   11.   Framing
_____   A. Basement walls   _____ _____
_____   B. 1st floor   _____ _____
_____   C. Walls-1st floor   _____ _____
_____   D. 2nd floor   _____ _____
_____   E. Walls-2nd floor   _____ _____
_____   F. Sheathing   _____ _____
_____   G. Ceiling joists   _____ _____
_____   H. Rafters/trussed   _____ _____
_____   I. Decking   _____ _____
_____   J. Felt   _____ _____
_____   12.   Roofing   _____ _____
_____   13.   Set Fireplace   _____ _____
_____   14.   Set doors   _____ _____
_____   15.   Set windows   _____ _____
_____   16.   Siding and cornice   _____ _____
_____   17.   **Rough Plumbing**   _____ _____
_____   18.   **Rough HVAC**   _____ _____
_____   19.   Rough Electrical   _____ _____
_____   20.   Measure cabinets   _____ _____
_____   21.   Rough phone/cable   _____ _____
_____   22.   Rough miscellaneous   _____ _____
_____   23.   Insulation   _____ _____
_____   24.   Framing inspection   _____ _____
_____   25.   Clean interior   _____ _____
_____   26.   Sheetrock   _____ _____
_____   27.   Clean interior   _____ _____
_____   28.   Stucco   _____ _____
_____   29.   Brick   _____ _____
_____   30.   Clean exterior   _____ _____
_____   31.   Septic tank   _____ _____

		DATE DONE:	NOTES:

_____	32.	Utility lines	
_____		A. Gas line	_____ _____
_____		B. Sewer line	_____ _____
_____		C. Phone line	_____ _____
_____		D. Electric line	_____ _____
_____	33.	Backfill	_____ _____
_____	34.	Pour garage	_____ _____
_____	35.	Drives and walks	_____ _____
_____	36.	Landscape	_____ _____
_____	37.	Gutters	_____ _____
_____	38.	Garage door	_____ _____
_____	39.	Paint/Stain	_____ _____
_____	40.	Cabinets	_____ _____
_____	41.	Trim	_____ _____
_____	42.	Tile	_____ _____
_____	43.	Rock	_____ _____
_____	44.	Hardwood floors	_____ _____
_____	45.	Wallpaper	_____ _____
_____	46.	**Plumbing (final)**	_____ _____
_____	47.	**HVAC (final)**	_____ _____
_____	48.	**Electrical (final)**	_____ _____
_____	49.	Mirrors	_____ _____
_____	50.	Clean interior	_____ _____
_____	51.	Vinyl/Carpet	_____ _____
_____	52.	Final trim	_____ _____
_____	53.	Bath Accessories	_____ _____
_____	54.	Paint touch up	_____ _____
_____	55.	Final touch up	_____ _____
_____	56.	Connect utilities	_____ _____
_____		DATE COMPLETED:	_____ _____

DRAW SCHEDULE SHEET

JOB DESCRIPTION _____ JOB NO. _____ BLDG. PERMIT NO. _____

NAME OF BANK _____ ADDRESS _____

LOAN PAYOFF .... _____

CONSTRUCTION LOAN CLOSING AMOUNT .... _____

DRAW	DESCRIPTION	DATE	AMOUNT
One	_____	_____	_____
Two	_____	_____	_____
Three	_____	_____	_____
Four	_____	_____	_____
Five	_____	_____	_____
Six	_____	_____	_____
Seven	_____	_____	_____
Eight	_____	_____	_____

TOTAL DRAW AMOUNT .... _____

TOTALS OF CLOSINGS AND DRAWS .... _____

TOTAL INTEREST PAID ON CONSTRUCTION LOAN .... _____

BUILDING CONTRACT

(date)_____

1. The undersigned owner authorized the undersigned builder to Construct and deliver to said owner a dwelling in accordance with the plans and specifications which are attached hereto and made a part hereof, on all that tract or parcel of land lying and being in Land:

Lot_____ of the _____ District
of_____ County,_____(state), and being known
_____

2. Except as proved in the paragraph dealing with allowances hereafter, the total cost to the owner (excluding the cost of the land and additional arranged work), shall be:

_____DOLLARS($_____)

Said amount shall be paid as follows:

SAMPLE BUILDING CONTRACT

3. Builder shall commence construction of the house in accordance with the attached plans and specifications after the construction loan has been closed. The construction of the house shall be completed within _____ days from the time that the construction is commenced.

4. The improvements shall be deemed fully completed when approved by the mortgagee for loan purposes.

5. Builder agrees to give possession of the improvements to the purchaser immediately after the full payment to the builder of any cash balance due. At that time, the builder agrees to furnish the purchaser a notarized affidavit that there are no outstanding liens for unpaid material or labor.

6. It is the understanding of the parties that the purchase price above described includes builder's allowances for items hereinafter described. The amounts are listed at builder's cost. It is the understanding of the parties that the purchase price shall be equal to the to the price described in Paragraph 2 above, provided the costs of the items below listed do not exceed the amounts shown. In the event that the builder's cost exceeds any item hereinafter listed, the purchase price shall be increased by the amount above the allowance shown. The allowances are as follows:

7. The cost of any alterations, additions, omissions or deviations at the request of the owner shall be added to the agreed purchase price. No such changes shall be made unless stipulated in writing and signed by both parties hereto.

8. Builder shall provide to owner, in writing, a twelve (12) month construction guarantee.

9. Every effort will be exerted on the part of the builder to complete the construction within the period described above. However, builder assumes no responsibilities for delays occasioned by causes beyond his control. In the event of a delay of construction caused by weather, purchaser's selections, etc..., the period of delay is to be added to the construction period allotted to builder.

10. This contract shall inure to the benefit of, and shall be binding upon, the parties hereto, their heirs, successors, administrators, executors and assigns.

11. This contract constitutes the sole and entire agreement between the parties hereto and no modification of this contract shall be binding unless attached hereto and signed by all parties to the agreement. No representation, promise, or inducement to this agreement no included in this contract shall be binding upon any party hereto.

The parties do set their hands and seals the and and year first above written.

_____
(Owner)

_____
(Owner)

_____
(Builder)

CONSTRUCTION STIPULATIONS

Special stipulations from sales contract dated_____ for the purchase of property located at _____.

1. The movement of any household goods or other materials by Purchasers into the house will not be permitted until house has been completed and accepted by Purchaser and the total sales price has been paid in full (closed).

2. Every effort will be exerted on the part of the seller to complete the construction of said house within the projected period of time, however, Seller assumes no liability for delays occasioned by causes beyond his control.

3. The costs of any alterations, additions, omissions or deviations at the request of the Purchasers shall be added or deducted from the agreed sales price, but no such changes shall be made unless stipulated in writing and signed by all parties, hereto. Any changes that increase the cost of completing construction will be paid for by the Purchaser prior to commencement of work.

4. Effective with the final acceptance and closing of transaction by the Buyer, Seller will provide a ten-year Home Owners Warranty with exception of anything the Buyer might have subcontracted on his own account.

5. The following standard features are included in the sales price and in the structure: (1) Natural gas heating and hot water (2) Smoke detector (3) Double stainless steel kitchen sink (4) Color co-ordinated gas range, vent hood, and automatic dishwasher (5) Fiberglass tub (6) Vanities in baths (7) Mirror in each bath (8) 8x10' patio (9) Central air.

6. House will be built per attached plans and specifications of materials marked Exhibit "C" and "D".

7. Seller stipulates that he will promptly correct any leakage or seepage of exterior surface water into basement of dwelling for a period of 12 months from the date this sale is closed. Seller will not be responsible for and will not correct any leakage or seepage caused by: (a) breakage or bursting of water mains or pipes, (b) any grading done by Purchaser which causes water to flow towards outside foundation wall, (c) prolonged direction of water against the outside foundation wall from water spigot, sprinklers, hose, broken gutters clogged with leaves or pinestraw, downspouts, broken, bent or clogged.

8. Decorator features such as color of carpet and appliances, may be chosen by the buyer from the price range and samples offered by suppliers chosen by the Seller. The following allowances are provided by the Seller. Buyer may select higher cost items and if so selected, agrees to pay the Seller's

suppliers difference in the allowance and the higher cost of items selected at time of selection:

a. Lighting fixtures including bulbs, chimes, tax _____ $_____
b. Wall paper per single roll not installed _____ $_____
c. Carpets per sq. yd. including tax, pad, and installation _____ $_____
d. Vinyl per sq. yd. including tax and installation _____ $_____
e. Foyer floor per sq. yd. including tax and installation _____ $_____
f. Landscaping allowance, if applicable _____ $_____
g. Appliances (where applicable):
   (includes disposal, dishwasher, oven with hood, and stove _____ $_____
h. _____ $_____
i. _____ $_____
j. _____ $_____
k. _____ $_____
l. _____ $_____

By: _____   _____
               SELLER   BUYER

By: _____   _____
               SELLER   BUYER

CHANGE ORDER

JOB:_____ DATE:_____

Whereas the parties hereto have made and entered a Contract and Agreement dated the _____ day of _____, 19_____, whereby the undersigned Contractor has contracted to build for the undersigned owners a house and other improvements in accordance with and pursuant to the plans and Specification of materials comprising said Contract;

AND WHEREAS the parties now desire to change, add to, and/or modify certain provisions of said Contract; the parties have reached agreement as to additional amounts to be paid Contractor by Owners for Contractor's agreement to make the changes, additions and/or modifications as requested by Owners;

NOW THEREFORE, the parties do hereby agree that changes and/or additions are to be made as follows in the provisions of the original contract and agree to pay to Contractor the following additional amounts for the making of such changes and/or additions, said amounts to be added to the basic contract price and to be paid Contractor by Owners pursuant to provisions in the original contract.

ADDITIONAL
CHANGE DESCRIPTION _____ CHARGE TO OWNERS

_____ _____

_____ _____

_____ _____

_____ _____

_____ _____

_____ _____

_____ _____

_____ _____

_____ _____

TOTAL _____

_____   _____
OWNER   CONTRACTOR

_____
OWNER

274 ■ THE COMPLETE GUIDE TO CONTRACTING YOUR HOME

Index